Fenella.
April 15th 1987.

THE RARE BREEDS HANDBOOK

Derek Wallis

BLANDFORD PRESS
POOLE   NEW YORK   SYDNEY

First published in the U.K. 1986 by
Blandford Press, Link House, West Street,
Poole, Dorset, BH15 1LL.

Distributed in the United States by
Sterling Publishing Co., Inc., 2 Park Avenue,
New York, N.Y. 10016.
Distributed in Australia by
Capricorn Link (Australia) Pty Ltd,
Po Box 665, Lane Cove, NSW 2066

British Library Cataloguing in Publication Data

Wallis, Derek
  The rare breeds handbook.
  1. Livestock breeds—Great Britain—History
  2. Rare animals—Great Britain—History
  I. Title
  636.08′21′0941   SF105

ISBN 0 7137 1615 0

Typeset in 10/12pt Garamond by
Asco Trade Typesetting Ltd., Hong Kong

Printed in Great Britain by
Garden City Press Ltd., Letchworth, Herts.

Contents

Foreword

BUCKINGHAM PALACE.

The judging of domestic livestock at agricultural shows is probably regarded by the uninitiated as some sort of mystic ritual. In fact, the business of comparing the conformation and performance of animals is a very important factor in breed development.

The animals we see today are the way they are as a result of cross-breeding between a number of native breeds so as to combine the most desirable qualities from each. Some of the founder breeds of our modern stock are still kept for commercial purposes but others are becoming increasingly rare. All these native breeds represent an important reserve of genetic material, and the Rare Breeds Survival Trust deserves great credit for its efforts to save them from extinction.

This book is a timely reminder of the value of these rare breeds and of the part they played in the history of the development of commercial livestock in many parts of the world. It is not a scientific treatise; it traces the outline of the development of modern British breeds from domestication to the present day. It includes a number of amusing anecdotes, both probable and apocryphal, about their often unusual history and a description of the current status of each surviving native rare breed.

1986.

Author's Note

In the text, 'Black' cattle refers to cattle as a species and is not a reference to colour. Where 'black' cattle are mentioned, this does denote colour.

Acknowledgements

My thanks go to my wife Margaret for her help and encouragement in this project and for her invaluable assistance both in typing and correcting the text.

I would like to thank all the owners who have given us permission to photograph their animals and my special thanks go to Gerry Lockley whose patience and enthusiasm have been invaluable in the production of the majority of photographs in the book.

I am also especially grateful to Bernard Lewis and the City of Sheffield Recreation Department, and to Michael Rosenberg of the Rare Breeds Survival Trust for his help in providing information for the writing of the book. I would also like to express my gratitude to Lawrence Alderson, technical consultant of the RBST, and to Dr Juliet Clutton-Brock for reading the text and for giving me the benefit of their wide knowledge on the subject of rare breeds.

Other photographs and assistance which I would like to acknowledge have been received from the following.

The Rare Breeds Survival Trust
   Limited
Dowager Countess of Tankerville
The Chillingham Wild Cattle
   Association Limited
Alan Searle
Marquis of Cholmondeley, GVCO, MC.
Bill McDermott
Fred Tyldesley
The Deane School, Bolton.
David Yarsley
Bill Stephen
Merseyside County Council
Kathy Smith
Alec Paris Publicity Limited
Howard Payton
Gaynor Morris
S.H. Urbanski
R.M. Baker
Frank Woolham

Alan Cooper
Gavin Large
Peter Copeland
Judy Walker
Shugborough Park Farm
Cartwright A.V. Services
Richard Wallis
Mike Wright
Frank Mosford
Iolo Owen
Barbara Platt
D. Mansell
Exmoor Pony Society
Hans Falk
Wensleydale Longwool Sheep
   Breeder's Association
Agnes Winter
Liverpool Daily Post & Echo
G. Hughes
J. Wrench
Bolton Evening News

Introduction
The Origins of Domestication

Up to the time of the Roman Empire, very little is known about the domestication of animals and the origins of farming. Nevertheless, archaeologists have managed to piece together most of the general steps which must have taken place.

Early man led a nomadic existence as a hunter-gatherer, moving around in search of food and water. This way of life continued until Neolithic times, during which primitive land usage for growing food crops was developed. This enabled groups of people to settle in one place for longer periods, until the land was exhausted, when they had to move on to new and more fertile areas.

Permanent settlement was the next step. With this began the domestication of animals in larger numbers than had been possible before. It is thought that after dogs, in western Asia, sheep and goats were the first animals to be kept, followed by pigs and cattle. These were all used for food. In different parts of the world, horses, camels and llamas were tamed and used as beasts of burden, and man's scope for utilising and adapting his environment was consequently greatly widened. Agricultural techniques became more sophisticated, and as a result the population of the

Some of the RBST reserve flock of North Ronaldsay sheep on the Trust's island of Linga Holm in the Orkneys.

settlements began to grow.

The first cattle to be domesticated were descended from the wild ox, or aurochs (*Bos primigenius*), which was a fierce, huge-horned animal. Breeding in captivity resulted in a reduction in size, aggressiveness and length of horn, although the latter still varied considerably. The Red Jungle Fowl was domesticated in India, and from it are descended all the various breeds of domestic poultry. Cats were kept, initially in Egypt, to control birds and rodents, and later as pets.

The British Isles were separated from the Continent of Europe by the breaking through of the English Channel, which was a consequence of rising sea levels at the end of the Ice Age. The final separation occurred about 8,000 years ago at the beginning of the Mesolithic period in Britain.

Cattle and pigs were the first domestic animals in Britain at the beginning of the Neolithic period, approximately 5,000 years ago. These were followed by sheep and goats as the forests were cleared by the first farmers.

1 The Roman Era

Before the time of Julius Caesar, Rome was a republic ruled by nobles, but Caesar made himself lord over all and governed like a general commanding an army. He was to become the founder of the Roman empire, and one of his most famous conquests was that of Gaul, a country which was mostly made up of present-day France. The inhabitants were of the same Celtic type as those who lived in Britain and with whom they carried on much trade, and when Caesar tried to overrun Gaul tribes from Britain went to the aid of their kinsmen across the Channel. Caesar decided to punish the Britons for their actions and led two expeditions to put down the war-like people who had fought his army. The first took place fifty-five years before the birth of Christ, but on this occasion Caesar under-estimated the number of soldiers which he would require for the action and was forced to return to Gaul to avoid heavy losses.

The next year he returned with a larger army and did not leave until he had defeated the Britons and forced them to pay tribute to Rome. Besides being a statesman and a warrior, Caesar was also one of the first famous chroniclers. He wrote a full account of his findings in Britain, and, from these writings we have the earliest full descriptions of Britain and the people who inhabited it. It is known that agriculture was flourishing in the south of the country.

In 9 B.C. Roman geographer Strabo noted that Britain exported grain and cattle to the continent. The early farmers must, therefore, have grown a surplus of agricultural products which they traded with tribes across the Channel. For nearly one hundred years after the invasion the Britons were left to themselves while the great Roman Empire which Caesar had formed became established.

Before Caesar's invasion, Britons in the south traded with Gaul and those in the western areas with Ireland. Dogs, trained both for hunting and tending livestock, animal hides, gold from the Welsh hills, and slaves formed the main part of this traffic. The slaves were usually prisoners taken in battle, or girls who were sold by their families. Such trade brought wealth and possessions in the form of jewellery, ornaments and livestock. In their villages people grew corn which was ground into flour, using stone handmills, or querns. They hunted some game and tended livestock. They spun wool into yarn and wove cloth which was very similar to a crude Scottish plaid.

The Roman invasion brought changes which affected the way of life of most of the population, especially in the southern part of Britain. Trade which hitherto had been relatively local began to expand, especially between other provinces of the Roman Empire and Britain. Animals formed a considerable part of these dealings and resulted in new breeds and types of livestock being introduced to replace or be cross-bred with indigenous stock.

Livestock husbandry had virtually taken the place of hunting, but agricultural animals had altered very little during the late Bronze and early Iron Age. A few tribes which inhabited remote areas, who were virtually unaffected by the new importations, retained their animals more or less unchanged, and because of this some of the animals to be seen in the rare breed collections today were saved. They are of great interest, not only to the general public, but also to geneticists and historians.

Cattle

At the time the Romans came, cattle varied in size according to the locality in which they were kept, and they were mainly black in colour with sharp, pointed horns. However, it seems likely that some of the animals from the Western Isles of Scotland were probably light brown or dun-coloured, very similar in appearance to present-day Highland cattle.

Black and dun Highland cattle at Cholmondeley Castle, Cheshire.

Cattle were used mainly as draught animals. Milk may have been produced at certain times of the year, but was as yet an occasional, rather than a regular, part of their owners' diet. Calves, and animals too old to work, provided meat. Present-day examples of early British cattle are the Kerry, the Dexter and the Welsh Black breeds which, although they differ in size, all retain certain common characteristics.

The origins of Britain's White Park cattle may lie with the Romans. White cattle are to be seen in a number of parks attached to stately homes. They are attractive animals, white in colour, with either black or red ears and muzzles and splashes of the same colour on their legs and neck. One widely accepted theory is that they were brought to Britain by the Romans. The Roman historian, Pliny, wrote that Roman bulls produced good milking cows. He went on to say that cheeses produced in the valley of the River Po were shipped to Rome. These 'lunar' cheeses (shaped like a full moon), as they were called, weighed up to 1,000 lb each, and the cheese-making art was passed on to the Britons.

Another Roman writer, Aelianus, praised cattle found in the southern part of Italy and notes that most of the animals were white in colour. This, he says, was due to their drinking the water of the River Crathis, as all animals turned white by so doing. Despite such fanciful theories, we know that the Romans had cattle of far superior quality to those which were to be found in Britain, and we know that a good proportion of them were white. Among modern Italian cattle, white or whitish animals with long horns and red or black muzzles and ears, are very conspicuous. It is reasonable to suggest that our white cattle and those which graze the fertile Po valley are of common ancestry.

It is likely that the Romans brought cattle from Italy to improve British stock and

A typical example of a White Park cow from the Chartley herd, at Forest House, Kelsall, 1985.

that these were the ancestors of the present White Park animals, although, as in most populations of animals, white individuals are produced from time to time and white cattle could possibly have been selected over the decades from the indigenous Black Celtic stock.

Remains of oxen from Roman sites show evidence of the emergence of larger beasts than commonly existed in the preceding centuries, but equally there was no overall improvement in the stature of cattle. The size of bone shows that oxen quite as small as the Celtic Shorthorn (introduced by Neolithic herdsmen from the Continent) were still abundant.

A sufficient number of bones of cattle have been recovered from archaeological sites to show a size range spanning several modern breeds. Some were smaller than the Kerry, for example Dexter size, and others were as large as Friesians. In addition the skulls of hornless cattle have been found and this discovery is particularly interesting in that the polled cattle were of two types. In the one the frontal crest is flatter than Galloway cattle, and the other has a prominent poll as in the Aberdeen Angus. These animals may have been the very early ancestors of the two Scottish breeds.

As far as size is concerned, the range shown by Roman cattle brings them well within that of modern cattle, although they do not show the same robust proportions that the latter may achieve.

Sheep

Sheep were still of the small, horned, Soay type, short-tailed and coarse-coated,

A Soay ewe with twin lambs. These are unchanged descendants of prehistoric sheep.

A Ryeland ewe with lamb.

although there is some evidence to suggest that selection had taken place, as some of the wool remains which have been found appear to have a higher percentage of white fibres in them.

The first Roman imports of sheep were probably white-faced and mainly hornless, with long tails, unlike the original British sheep. It seems likely that they were used to cross-breed with the indigenous Soay type to produce offspring with a more acceptable fleece and yet still be capable of living on rough pasture. The pure Roman animals would have been unlikely to thrive under the harsher climatic conditions in Britain.

Some of the present-day hill breeds which are found in the west of Britain, such as the Welsh Mountain and the Cheviot, seem to owe their ancestry to this early breeding programme.

The medieval, hornless, short-woolled Merino is thought to have played a major part in the Roman improvement plan. A present-day example of this type of animal would be the Ryeland breed which all but died out and was rescued due to the efforts of the Rare Breed Survival Trust.

Cotswold sheep are thought to be one of the oldest breeds on record and it would seem feasible to suggest that they are one of the modern representatives of the first Roman importation of long-woolled sheep.

Other breeds which have been developed in certain areas for specific local needs are the Lincolnshire Longwool, Leicester Longwool and the Devon Longwool. The two former breeds were later to be used by Robert Bakewell in his sheep improvement programme.

A Cotswold ewe with lamb. This breed may be a direct descendant of Roman importations.

On examination of wool fibres from Bronze Age textiles, it was at first thought they must have been mixed with deer hair but it is now realised that the fibres so described were really the kemp that forms the outer coat of the fleece of the wild sheep, and are still common in the fleeces of primitive domestic sheep today, of which the Soay is a good example.

Fleeces had improved by the Roman Iron Age. Fewer hairs have been found in the cloth from this period. In addition, whereas the Bronze Age wool was mainly brown, there were more white fibres in the Iron Age wool. It is known that the Romans had a well organised wool-textile industry in Britain and there is the classical reference of Diocletian, circa A.D. 300, to the British wool 'so fine that it was comparable to a spider's web'. Among several specimens of Roman textiles taken from Scotland and examined microscopically was a white, true, fine wool that justified this description.

Wool, to the Romans, was a valuable commodity and they would not have been very impressed with the early British sheep. Their imported breeds seem to have brought about a great improvement in fleece quality and this resulted in the setting up of an extensive cloth-making industry in certain areas of the country. One such is known to have been in Gloucestershire and it is thought to have given the name 'Cotswolds' to that district; 'cote' was the name given to a special building such as a pig-cote or pigeon-cote, in this instance a sheep-cote, and 'wolds' refers to a wild, open hilly area, such as the Lincolnshire wolds. This suggests that the area itself was, therefore, named after the sheep, not vice versa.

Goats

Goats were the second type of animal to be domesticated by man in western Asia. They had been herded for more than 7,000 years by the time of the Roman invasion, but the Romans developed a good milking strain in other parts of their empire and these, too, were brought to Britain in an effort to improve the resident stock. Many present-day goats have a pair of fleshy growths on the neck, which hang down rather like toggles. It is said this characteristic was introduced into Britain by the Romans. Roman herd-masters were instructed to use billys, or male goats, with this special feature. It resulted in the daughters of the improved animals having toggles and thus made it possible to see at a glance which goats contained imported, and therefore improved, blood. This form of selection, helped by an easily recognisable characteristic, played a major part in the breeding policy adopted by the Romans.

Pigs

Some experts are of the opinion that the invaders also brought with them partially domesticated pigs which had been developed in the eastern part of their empire. They differed in shape from the semi-wild British animals and probably had a less hairy coat, due to their origins in a warmer climate. They almost certainly had flop ears. It is a characteristic said by some people to have been passed on to many present-day breeds of pigs, but this seems to be questionable. The Roman animals would have taken far more readily to management, being more docile than the

Early Chinese ancestors of British breeds of domestic pigs.

Early Asiatic ancestors of British breeds of domestic pigs.

A British Saddleback sow. This is a combination of the Wessex and Essex Saddleback breeds.

British pigs of the time. Modern breeds which carry this flop-eared characteristic include the Saddleback, the Gloucester Old Spot and the Large Black, but all these breeds seem to have been developed from much later imports into Britain.

Horses and Ponies

The Romans found ponies the size of our modern Exmoor/Shetland pony too small for their needs when they landed in Britain. They therefore introduced a bigger type of horse for use mainly as a pack or riding animal. Remains of Roman horses which have been found seem to suggest the size to be somewhere between thirteen and fourteen hands (a hand being four inches). This would be about eight inches taller than the indigenous animals.

The main draught animal was the ox, due mainly to the fact that at this time the horse collar was unknown. Oxen were used to pull heavy loads owing to the ease with which they could be yoked. The horse's strength is in its shoulders and it is only with the advent of the collar that horsepower could be fully utilised. The Romans used stallions to pull their chariots and, though these animals were swift and powerful, having developed strong muscles, even they would not have been capable of pulling heavy loads over long distances. The Roman harness consisted of a crest-plate and yoke, and this limited the use which could be made of horses.

An Exmoor pony and foals on their native heath.

Nevertheless the breed was improved by the addition of new blood during the Roman era and the horses gradually increased in size and form.

By the beginning of the Christian era, there are signs of extensive horse-breeding in Britain. A find, dated between 100 B.C. and A.D. 50 in Anglesey, contained many bridle bits. The domesticated horse was now playing a leading part in British life which it continued to do for the next 2,000 years. During the Roman occupation, it was an animal for military purposes, conferring an immense superiority in mobility and offensive warfare upon its rider or driver, and it seems that it was not taken into general use as an agricultural draught animal in Britain until the Roman occupation at least.

Poultry and Other Animals

Roman imports of poultry and pheasants were to add to the delights of the dinner table among the wealthy. The ring-neck pheasant was first introduced into Britain by the Roman army. Though these birds are still to be seen throughout the country, the population is largely maintained by artificial hatching and rearing of vast numbers of birds each year, which are destined to provide targets for sportsmen.

Another delicacy which Roman gourmets brought to Britain, was the fat-tailed or edible dormouse. These animals were fattened in stone urns and fed on cereal, fruit and nuts. Breeding colonies were set up to ensure a constant supply for the table. In parts of the south of England there are still areas where the edible dormouse thrives and these animals are direct descendants of those brought here by the invaders.

The Romans brought civilisation and peace to Britain. They left a great inheritance in the form of livestock improvement. They left these islands after some 350 years of occupation, years which to this day affect our lives. There are many examples of Roman architecture in Britain to remind us of their building skills, and we are even more fortunate in having some living examples of the positive contribution they made to our emerging agricultural system.

The Romans left Britain when conditions in their homeland made it impossible to sustain an empire. Our ancestors were thus left to look after their own affairs, but the conquerors' influence over so many years meant that they were totally unable to organise themselves. Without Roman help, many aspects of life in Britain suffered, and agriculture was one of them.

2   The Viking and Saxon Era

The departure of the Roman legions in the first half of the fifth century heralded a period in the history of Britain known as the 'Dark Ages'. The six hundred years until the Norman accession took British agriculture and livestock improvement into eclipse.

Place names such as Shipley (Sheepfield) and Skipton (Sheeptown), are derived from this time and give some clues to historians. Both Celtic and Saxon words, which are retained in town and village names, point to the area being the centre for a specific trade or occupation. The *Anglo-Saxon Chronicle* also refers to the depredation by disease or marauders upon livestock at that time.

After the Roman departure, the country was looked upon as easy prey by war-like Celtic tribes from Scotland and Ireland. Our ancestors had grown accustomed to Roman, or as it became later, Romano-British rule. The army protected the civil population and agriculture prospered in a relatively law-abiding period. Stock depradation caused by theft was uncommon and the protection afforded by the state enabled the majority of the agricultural population to concentrate on their livestock and on their farm husbandry.

With the breakdown of the Roman order, great changes took place. Pleas were sent to Rome for help and for a short period soldiers returned to assist the Britons in defending themselves from these attacks, but they were again recalled to Rome and Britain was left to its own devices. Raiders from around the River Elbe in Germany invaded the shores of this country. Their attacks devastated the coastal settlements first, but later the whole of the south-east of the country was over-run. Known as Jutes, these raiders did not believe in Christianity but worshipped their own pagan gods. They eventually formed the kingdom of Kent and also took over the Isle of Wight. The Saxons from northern France and Holland developed their own type of agriculture and built villages along the south coast of Britain. They formed their own settlements of Sussex (South Saxon), Wessex (West Saxon), Essex (East Saxon) and Middlesex (Middle Saxon), whilst the third tribe from Europe, the Angles, colonised the rest of the country, which included East Anglia (East Angles) and Mercia, the heart of the country, and Northumbria which stretched to the Scottish border.

The Britons were driven to the west and north of the country, and although they remained in areas we now know as Cornwall, Wales and the Lake District, and the western strip of Scotland, the rest of the country eventually became one kingdom under the overlordship of Egbert, the Saxon king of Wessex. Later, the Vikings from Scandinavia sent raiding parties to Britain and over the following years they formed settlements and eventually became permanent residents.

It is justifiable to suppose that the immigrants to these islands brought with them some animals. Even if these were intended to be slaughtered for food, some individuals are sure to have lived and bred, adding yet more blood to the stock which already existed.

Records of the period between A.D. 800 and 900 list farm livestock as oxen for work, and cows which are thought to have been kept to produce working oxen and beef calves, and in a few cases provide milk for human consumption. Wethers (sheep) were also listed. This scant information gives no details of the type or colour of livestock.

Wills often referred to areas of woodland together with herds of swine, in some cases up to 2,000 in number. Records also give account of deaths of livestock from 'severe winters and the attacks of heathen men'. 'What the marauders left, the act of God took' was recorded in A.D. 896.

Pigs

Pigs seem to have been ubiquitous in Saxon England and to have provided a major occupation in the countryside, as the production of pork was of great importance. Whether they were 'yarded' as a general practice is a matter for speculation, but it would seem reasonable to suggest that at least part of the life of the Saxon pig was spent housed in a sort of 'hog-yard'. The damage created by vast herds of pigs on undrained land is not difficult to envisage and yards would also have made the selection of stock much more simple.

A 'modern' in-pig Iron Age sow, constructed by Joe Henson by crossing a wild boar with a Tamworth sow.

Little information as to the type of pig is available, but it was descended from *Sus scrofa*, the wild boar, long-legged, long-tailed, razor-backed and slow maturing. It was very little changed from its Iron Age ancestor, a forest grazer which produced lean meat by the time it reached three years of age, or older. If the Romans had introduced a better type of animal with short legs and less hair which was also tamer and more easy to control, it had probably by this time degenerated or died out, as the type of husbandry together with the climate would not have been suitable for such imported stock.

The pig reached its peak in pastoral areas during Saxon times, when it lived off forest produce in the autumn and winter, and grass during the spring and summer. In arable areas of the country, the pig played little or no part in the economy and this points to the fact that it was an animal which was expected to find its own food. It was a totally different animal to our modern pig, which has been developed to produce and grow in an almost completely controlled environment.

Cattle

During this period, due to the settlers from different areas of Europe and Scandinavia bringing their own type of livestock, especially cattle, to Britain, we see the development of localised strains rather than breeds. Each type of animal would have been developed over many years and probably have characteristics unlike those of animals from different locations. Therefore, where the invaders settled in this country, the cattle which they had brought with them eventually became the local 'breed' or developed into the local strain, whereas the original British cattle are thought to have survived only in remote parts of the country such as Wales, the west of Scotland and Ireland, where they were hardly affected by the new introductions.

With the introduced livestock, as well as the indigenous animals, there seems to have been a wide variety of cattle in Britain, from the large-framed oxen used by the Saxons on the arable farms in the south of the country, to the small Iron Age type like the modern Dexter or Kerry, which was to be found on the west coast.

During the eleventh century, the horse collar was introduced from France, but it seemed to have had little effect on the use of oxen as the main agricultural draught animal.

Cows kept especially for milk production were still thought to have been very rare at this time, but could possibly have been kept on small farms. The smaller farms may also have used the same animal to supply traction, a calf and a little milk, the latter being purely a by-product obtained for a few weeks in the spring or early summer, and would have amounted to perhaps only about a pint a day as the calf would also have needed to receive its share.

As there are no known records of cattle being kept primarily as dairy animals, there can be no proof; but it seems likely that on a very small scale the dairy cow had begun to appear on the scene. Generally speaking, it is accepted that milk during this period was produced almost exclusively by sheep and goats, while cattle provided draught animals which at the end of their working lives, were fattened and slaughtered for beef, and surplus calves which were also used for meat. There is some evidence of bullocks having been both stalled and grazed specifically for meat production. This seems to bear out the theory that different breeds were being developed for specific purposes. Some strains of animals would have been found to fatten more easily than others, and these could have been used specifically for meat production.

The cattle in the south and the east of the country appear to have been of the improved type and were at least as good as Roman cattle, but to the north and the west they were thought to have degenerated into a smaller type altogether.

By the year A.D. 900 the Vikings had taken control of and therefore had influenced much of northern Britain. The process had been a gradual affair and with the inter-breeding of the new types of animals and those already prevalent in the area, yet more types were developed. One characteristic, once thought to have been introduced by the invaders, was the hornless, or polled, factor in cattle. This has caused much argument between historians over the years, as the remains of polled cattle have been discovered on Roman and Iron Age sites, whereas the remains of cattle on known Viking sites, such as those on Orkney or the Shetland Isles, have turned out to be from both polled and horned animals. In an article written in 1950 entitled *Polled Cattle* Dr John Hammond points out that the polled or hornless condition in cattle arises as a spontaneous mutant once in every 50,000 births and this could account for their appearance throughout history. It seems likely, however, that the Vikings may have selected a hornless strain of cattle which would have proved far more easy to transport than animals with horns, on their long sea voyages. The polled gene is dominant over the horned, and it would not have been too difficult to amass a nucleus of hornless animals for this specific purpose. The ancestors of the Red Polled breed may have been developed at this period.

Sheep

During Saxon times, sheep appear to have been very widespread throughout the country and in the area which is now England it became an animal of increasing importance. Wool was exported from Britain to France and Scandinavia and the cloths made at such places as Skipton (Sheeptown) were said to have been of the highest quality available.

The sheep at this time were beginning to take on two more specific roles, that of producing milk and also that of land improvers. The latter role appeared to be the first sign of the appreciation of the 'golden hoof', this being the increase in the fertility of land brought about by the sheep's droppings and its improved soil structure due to the sheep's pattern of grazing.

The marshlands of Kent and the Thames estuary are thought to have been the centres of large-scale dairying based on sheep. Old writings mention these areas as being also important for the production of fat lambs and for cheese. It seems possible that the introduction of Saxon sheep brought about the black-faced characteristic which is now to be found in many of our present-day breeds, for example the Scottish Blackface or Swaledale.

By this time local types of sheep had developed along specialised lines and through selection and close-breeding (mating two closely-related animals), albeit quite by chance, regional strains had evolved. The Norfolk horned breed, which was of great importance all through the Middle Ages, is thought to have originated during the Saxon era. Early writings speak of the Breckland sheep of Norfolk as being thrifty and able to survive on poor quality grassland. This breed, which is thought by some people to have been developed from the Roman Longwool, was famous for its dual-purpose role of milk and wool production. The milk was processed to make cheese and was sold in the cheese towns of Norwich and Ipswich. *Wich* in Old English meant 'cheese' or 'salt'. The Norfolk Horn was not a meat-

A Swaledale ewe with lamb. This is a popular hill sheep in the north of England, and one of Britain's hardiest breeds.

producing breed and was only fattened at the end of its productive life. As a pure breed it fell out of favour as farmland became enclosed, but survived in small pockets until a few years ago. The Rare Breeds Survival Trust has reconstructed the breed by using closely-related strains of sheep, but the last surviving pure Norfolk Horn died in 1977.

The Vikings, who brought their own breeds of sheep to the north and west coast of Scotland and settled as far south as the Lake District and the Isle of Man, developed a totally different type of animal. Prior to the Viking invasion, the naturally short-tailed, horned, hairy sheep (very little improved from its Iron Age ancestor) was the only one in these areas. The Roman, and later Saxon, influence had little effect on these animals and it was left to the Scandinavian breeds to bring about any change.

The Manx Loghtan is one such breed which is thought to be of direct descent from the Viking animals, being brown in colour and having short wool and a short tail. The Hebridean, although black in colour, has the same short wool and tail. Both breeds are multi-horned and are capable of having one, two or even three pairs, which grow to a length of up to 45 cm (18 in).

A Manx Loghtan ewe with single lamb. The word 'Loghtan' derives from the Manx words *lugh* (mouse) and *dhoan* (brown). It is a very primitive breed.

A Hebridean ewe. The breed was formerly known as the 'St Kilda'. This is another primitive, multi-horned breed.

Having been isolated on the Hebridean Islands and the Isle of Man, these sheep have changed very little in the last 1500 years. Although in some cases other breeds have now been used to cross with them, both the Manx and the Hebridean are of great historical interest and can be seen in rare breed collections.

As the true Soay and the Viking breeds were driven to the coast and offshore islands, the influence of the Saxon breeds began to take effect on the animals of mainland Scotland. The mixing of the Viking, Soay and Saxon blood may have seen the emergence of yet another type of sheep which was selected for the specialised wool production by the crofters.

The Orkney, or North Ronaldsay breed, could have been influenced by this selection. The purebred North Ronaldsay is one of the most interesting of all farm livestock, as it has over the years developed along very specialised lines.

A working party repairing the stone fencing on the island of Linga Holm in the Orkneys. The island is owned by the RBST and is the home of the Trust's reserve flock of North Ronaldsay (seaweed-eating) sheep.

Originally the people on the island built a seawall to prevent the sheep from treading and soiling the seaweed which was used in the production of iodine. With the introduction of new techniques, both in the production of iodine and in farming, the sheep were driven from behind the seawall and forced to survive on the seashore where they developed the habit of eating and surviving on a diet consisting almost entirely of seaweed. There is more about this on p. 122.

The Shetland breed, too, has survived to the present day and produces very fine wool which is still important to the croft industry of those islands. The breed has retained the primitive habit of shedding its fleece, just like its Soay ancestor. Unlike the latter, it has developed a range of colours from brown to grey and white.

Old-type Shetland sheep photographed at the end of the nineteenth century.

Horses

Like cattle, pigs, sheep and goats, horses at this time are sure to have varied from area to area both in size, shape and colour. The horse is thought to have been valued as a riding and pack animal and was known to have been of great importance for use in warfare. There is no evidence of its having been used to any great extent as a working horse until the introduction of the collar in the late eleventh century.

During the Saxon and Viking era, the people of the British Isles experienced great changes and, at least during the early years of the period, suffered depredations as they and their livestock were driven out of their homes and settlements by the many invaders to our shores. During these 600 years it is thought that the birth of livestock breeds as we know them began to take place. The regional strains of farm animals took their first steps and the ancestors of many of these old breeds are still to be seen in farm parks.

3 The Norman Conquest to the House of Tudor

On the eve of the Norman Conquest, England was a kingdom divided into shires and the people were of three classes, nobles, commoners and slaves. The latter had no rights and were sold like cattle in the market place. They were usually prisoners taken in battle, debtors or criminals. Most crimes at this time could be atoned for at a price. For example, a leg was valued at fifty shillings, and an eye or a toe at eleven shillings. A male slave cost £1, a good horse thirty shillings, and a cow or an ox five shillings, a sheep or a swine a shilling, and a goat two pence. A day's wages was about one penny.

Goats, sheep and cattle were kept for milk production. Oxen, in teams of four, were used to plough strips of land on the one common, unenclosed field in the village. Mills for grinding corn were driven by water or slave labour and most people kept a hive of bees, as honey was the only form of sweetening.

William the Conqueror is said to have brought the Great Horse to Britain and stallions were used by knights in armour. They were strong animals able to carry total weights of up to 400 lb, and are said to be the earliest ancestor of the famous Shire horses of England.

A typical Shire mare.

Due to the many conquests of Britain by invaders from different parts of Europe and Scandinavia, and the constant shift of local populations, the make-up of domestic livestock is sure to have been constantly changing during this period. As different types of cattle, sheep and horses were introduced, selected and cross-bred, some improvement is thought to have taken place.

The Domesday Survey of 1086 suggests that in every county were many teams of working oxen. For example in Norfolk there were 5,014 plough teams with up to eight oxen in each, and 2,107 other beasts, the latter were probably for producing replacements for the plough teams, but also supplying a little surplus milk for human consumption. Sheep were ubiquitous and, along with goats, were used mainly for producing milk. Wool and manure were looked upon as an important by-product. Flocks up to 2,000 in number were kept in such areas as Norfolk.

Pigs, on the other hand, were looked upon as the poor man's animal and little account has been taken of them in the Domesday Book. In certain areas large herds were still to be found, although generally speaking, after the popularity of the pig reached its peak during Saxon times, its number dropped dramatically. Horses were still not of much importance agriculturally, although the harrowing horse is mentioned in the Domesday Book. As a rule, it appears, at this time horses were bred for use as riding animals and stud farms produced them almost entirely for this purpose. Light tasks on the farms could have been carried out by horses, but this would have been an exception rather than the rule.

A South Devon cow, Britain's heaviest indigenous breed.

Little is known about the type of animals which were to be found in Britain during Norman times, but in different areas of the country the livestock would have been initially that which was brought by the foreigners who settled in the district in question. They would also have been inter-bred with the original livestock of that area. From this basis a certain amount of selection is sure to have taken place, leading to localised strains of animals being developed for specific purposes. Environmental pressures such as the husbandry, type of land, rainfall, etc. are sure to have influenced the farm livestock in each region. One example is the cattle in north-west Devon which are said by some experts to have been the fore-runners of the Red Devon breed. The Ryeland sheep, too, is said to have first appeared in south-west Herefordshire at about this time and, as its name suggests, was bred on the rye-growing land of the Welsh border.

Goats

When Richard the Lionheart was king he spent little time in England and only visited it twice during his reign of about ten years. On one of his visits he is said to have brought back to this country a herd of beautiful goats. Returning from his Crusades in the late twelfth century, as he crossed Europe he was given a present of a number of black and white, long-haired goats. These animals are thought to have been originally from the Rhone Valley, as the wild Schwarzhal goats which still inhabit that area look very like the animals which the Lionheart brought back to stock one of his royal parks. There they became beasts of the chase and are said to have remained until the reign of Richard II who, in the late 1390s, is said to have enjoyed a good day's hunting with Sir John Bagot on his Blithfield estate in Staffordshire. In appreciation, the king gave the royal herd of goats to his host and they became known from that day on as Bagot goats, and formed part of the family crest (see p. 133).

Bagot goats, which can still be seen in farm parks and collections in different parts of the country, remained in the park at Bliffield until 1975 when what remained of the estate was sold and flooded to form a reservoir. Some of the goats were donated to the Rare Breeds Survival Trust and have now been saved from extinction.

Cattle

During the reign of Henry III (1216–1272) the famous herds of park cattle were formed. The Dynevor herd of white cattle is thought to have originated in A.D. 800 but more recent information seems to date them about the time of Henry III. From the earliest times some white cattle with coloured ears had existed in Wales, along with the native black cattle. Many stories as to their origin have been told and once again where folklore and supposition end and fact begins, it is difficult to determine (see p. 106).

The Welsh cattle, when first mentioned, are described as having red ears, not black, and they seem to have been fairly well distributed in Pembrokeshire and Carmarthenshire, the borders of the Bristol Channel and the Isle of Anglesey. Little or no evidence has been found of their inhabiting the mountainous areas of the principality. The white cattle seem to have only been kept on the better, lower land.

G. Kenneth Whitehead, in his famous book *The Ancient White Cattle of Britain* written in 1953 says the Welsh cattle with red ears were first mentioned in the *Venedotian Code of Laws* ascribed to Howel Dda in the tenth or eleventh century, and the present of four hundred cows and a bull which Maud, wife of William de Braose, made to the queen of King John is recorded. At this time, it appears that the herds of White cattle in Wales were mainly a domesticated strain and had originated from the feral cattle of England and Scotland, as either gifts or paid in the way of fines (see p. 106).

The most famous of all the herds of White Cattle is the internationally known Chillingham Herd, which was enclosed during the reign of Henry III, when he gave consent for Chillingham Castle to be castellated and crenellated and a park wall built. At this time feral cattle roamed wild in the old Caledonian forests and a

Chillingham cattle, still to be seen in their Northumberland home. The feral herd is still wild and completely independent of man.

number of animals were possibly driven into the 1100-acre park, and together with red and fallow deer were used by the powerful Norman family, the Grays of Wark-on-Tweed, as beasts of chase. There is evidence of the park's existence in 1292 and also in an Inquisition of Edward III in 1369 there contains a reference to a park with wild animals (see p. 114).

The Chartley Herd started in much the same way as the Chillingham Herd. After the Norman Conquest, Chartley Park passed into the hands of a powerful Norman baron, Ferrers of Chartley, and he like other landowners of the time, enjoyed what are today known as field sports, one of the most popular of which was hunting.

The feral cattle, together with other game species of the Forest of Needwood, which covered most of Staffordshire at that time, were by the king's permission driven into the emparked Chartley estate, there to remain until 1904 (see p. 107).

Other herds, in all probability, were formed at this time, as the large estate owners enclosed vast tracts of land for the sole purpose of retaining game in close proximity to their homes. The Cadzow Herd, the Cambenall Herd and the Vale Royal Herd are typical examples.

Social Changes

During Henry III's reign towns were beginning to develop; mayors were appointed and the first merchants' and crafts' guilds were formed. Street names were derived from the trade or industry which took place there: for instance, Bread Street and Wine Street. Those who worked on the land were still treated as slaves in most cases. Wages had increased slightly to two pence a day for men and one penny a day for women, and a whole family could be sold by their master in the same way as he sold his livestock.

In 1349 a plague called the Black Death spread misery all over Europe, and in England it is said that as many as one man in every three died from it. The large estates owned by the church and manorial lords took over the livestock of their tenants as they died or fled the law, and evidence of the different classes of farm livestock is available from records which were kept at the time. Plough horses, cart-horses, cows, steers and calves, wethers, ewes and lambs, pigs and piglets, and hens, are all mentioned in these records.

Country-dwellers lived off their tenanted land and were able to sell or barter any surplus produce which was left over after they had paid their dues to the manor. Single cattle or complete herds were often confiscated or taken in fines by the king. Markets were developing for the sale of surplus produce at this time and in the year 1386, nine years after the accession of Richard II, Leicester Abbey is recorded as having bought at Loughborough market three oxen and a bull, and ten oxen from Derby market.

Advances in Farming

Movement of livestock about the country was beginning to take place, but there are sure to have been small pockets of livestock which were unaffected by this movement and remained unchanged. The Rhiw breed of sheep from the Lleyn peninsula and the Polled Welsh Black cattle from the hills above Dolgellau (Dolgelly) were, until they died out a year or two ago, typical examples (see p. 60).

Welsh Mountain sheep on Rhiw Hill in the Lleyn Peninsula of North Wales. Their ancestors, the Rhiw breed, became extinct in the last decade.

A Welsh Mountain yearling ram.

Wool remained a valuable by-product of the milking ewes and at about this time rose once again to become of major importance to the sheep farmers. Lincolnshire and Cotswold rams were greatly valued for the quality of their wool and were sent further and further afield to improve the flocks of the landowners.

Canterbury sheep, possibly the ancestors of the Romney Marsh breed, are also reputed to have been producers of fine wool.

A red bull, taken from a tenant, is recorded as having come into the possession of Tavistock Abbey in 1366, and may point to the fact that a red breed of cattle was developing in that area, and that this particular bull was of some importance.

The revolt of the peasants took place in 1381 and marked the end of villeinage, whereby taxes were paid in the form of forced labour to the lords of the manor. From the thirteenth century onwards cows became more important as producers of milk, the raw product for the production of butter and cheese. The cow milk was mixed with sheep's milk where the two forms of stock were kept together. Young animals were often reared on the skim which remained after the cream had been removed.

With more movement about the country of agricultural products, certain areas were beginning to become well-known for the quality of their produce, and it seems reasonable to suggest that selective breeding of livestock took place on the large farms and estates owned by the church or landed families, and breeds or localised strains of animals from these areas took on the name of the locality. Some of our old breeds, such as the Gloucester cattle, survived in large numbers until the seventeenth century. Since then, especially during the last hundred years, they have only managed to survive with the help of a few enthusiasts. In 1970, for instance, there were only fifty pure-bred Gloucester cows in one herd left in Britain, but, due to the efforts of the Rare Breeds Survival Trust, they can now be seen in a number of private collections about the country.

Pigs

With the extension of ploughed land and the felling of the forests during Saxon times, pig numbers dropped considerably in the country, but a new form of husbandry had started to develop. In order to produce meat at times other than in the winter months, some pigs were kept in yards and fattened. Nursing sows were also housed in pig sties at certain times of year. Up to this time the pig had changed very little in either type or temperament, so it again seems likely that some selection would have taken place to develop strains which were more amenable to the new form of husbandry. Possibly, once again, local strains arrived on the scene. Although there is no real evidence of the importation of a different type of pig, no doubt some animals from Europe would have been used to cross-breed with local animals.

Horses

By this time bigger horses had been developed and were used mainly to carry knights in armour, but it would seem likely that they had also been developed for other purposes. Maybe a horse replaced some oxen in the ploughing team and there is little doubt they would also have been used to pull carts or other agricultural implements. As horse breeding and selection took place, they too, like other farm animals, took the name of the localities where they were developed—The Suffolk Punch, the Clydesdale from Scotland and the Shire from the Shires of England.

By the beginning of Henry V's reign the towns and cities of England had grown and the population of London alone was over 50,000. The people who lived in the town brought trade to the farms and estates in the surrounding areas and this in turn brought about yet more change in the type of agriculture carried on in the more highly populated areas of the country. During his reign gunpowder was first used and this put an end to the use of the horse as a weapon of war, at least as the mount of a knight in armour. The Great Horse was relegated to its role of hauling vehicles.

During the last half of the fifteenth century, the trade guilds extended and grew in number, but still the poor people in the countryside suffered at the hands of the landowners. Nobles and dignitaries of the church lived in great luxury. For example, at a feast given by the Archbishop of York the following were eaten; 400 swans, 400 plovers, 500 partridges, 4,000 pigeons, 500 stags, 4,000 venison patés, 1,500 hot patés, 200 hot custards, 12 porpoises and 1,000 dishes of jelly.

From the year 1066 until the crowning of Henry VII, the first King in the House of Tudor, in the year 1485, the changes which had taken place in England had been very dramatic. Farm livestock had begun to develop still further into local breeds, or strains. This process had taken the best part of 500 years. Animals which had been

An engraving of a team of oxen harrowing-in seed corn.

left when the Romans departed the country were cross-bred with animals which had been brought here by raiders from different parts of Europe and Scandinavia. Wherever the invaders settled their own stock predominated. The remnants of the feral cattle had been placed in parks and some of these herds remained in an almost unchanged form until today (see p. 106).

Towns and cities had expanded and the market for agricultural products had begun to become more specialised. The horse, in the latter part of the period, was given some of the work which had hitherto been carried out by oxen.

4 The Sixteenth and Seventeenth Centuries

Henry VII founded a strong dynasty and brought peace and order to England. He regulated the price of wool and prevented the export of gold. The population of England and Wales had dropped by 200,000 during the fifteenth century and totalled no more than $2\frac{1}{2}$ million at the beginning of the Tudor period. The rich had a passion for good living and records show they ate boar's head, beef, mutton, pork, swans, chicken, wildfowl, rabbits and game. This period heralded yet another agricultural revolution, the change from subsistence farming to a new form of commercial agriculture. Though the new era saw the development of still more specialised and localised strains of farm animals, true breeds did not begin to appear until the latter part of the seventeenth century. The change from subsistence farming was brought about by the rapid build-up of the towns and led to more and more people being unable to, or not wanting to, keep their own livestock.

As other trades and industries developed, so the production of food which the workers from this fast-growing sector of the country required, had to be met by the receding farming population. The principle of each family growing its own food and any surplus being sold or bartered for other necessities changed during this period. As the town and city population required more and more food, the need for greater production of red meat and other livestock products became all important.

With the release of the great monastic estates in Henry VIII's time due to the dissolution of the religious houses, the breakdown of the feudal system of land tenure accelerated. Other aspects also favoured the change in the new agricultural system; the peasant families found employment in the towns and moved from the country districts. So a new class of country-dweller took on the role of farming. These were businessmen turned farmers and they brought about a whole new dimension to the agricultural scene. The peasant arts of soil cultivation and farm livestock management were transformed into a new industry. The traditional methods of management, together with the still virtually unimproved specialist types of livestock, were not now up to the required standard. Types of cattle, sheep and pigs and, to a much greater degree, horses called for a change in emphasis. The major factor was not now just the production of food, but the profit which could be made by selling that food to the ever-growing market.

Wool remained an important cash crop, but meat, butter, cheese and even milk were required in increasing quantities by the hungry mouths of the town-dweller.

The Role of Livestock

The emphasis of production of farm livestock took a change of direction during the two hundred years of the Tudor and Stuart era. The cow took on a new role as a producer of meat and milk, as its importance as a supplier of oxen for draught use lessened. Sheep assumed a greater importance as meat producers as well as providers of fertility for the ever-increasing arable acres, though the quality and quantity of their wool remained paramount.

Horses were beginning to take on a new role, that of replacing some oxen in the plough team, as well as claiming their new place on the battlefield and on the highway. They needed to be bred with more strength and power.

The pig progressed slowly into a more domesticated farm animal, but was still basically the old type of forest animal.

During Henry VIII's reign sport took precedence over agriculture. Deer ran wild, ruining crops, and oxen were still used in the main arable areas for ploughing. Henry VIII also took a great interest in the navy, and salted beef became a most important part of the diet of sailors; this market was exploited by livestock breeders near to the ports.

Both home and overseas trade prospered. Inland, rivers were used for transporting agricultural products whereas livestock such as cattle, sheep, turkeys and geese were driven along the highways to the big markets such as Smithfield in London. Often livestock covered two or three hundred miles in this way.

During the early Tudor period, vast amounts of money could be made out of wool and this attracted the wrong type of people into agriculture. Peasants were dispossessed of their land which was incorporated to form large sheep runs, much like the ones in the Cotswolds which were developed during the Roman occupation. As the land passed into hands of speculators, the country swarmed with unemployed. The price of wheat rose and became too expensive to buy for the many starving people who begged in the towns. Severe penalties were administered on beggars who were whipped or put to death. Plagues of fleas, rats and mice were common and illness was rife.

During the reign of Queen Mary, travel in coaches became more common and people moved about the countryside in this first form of mass travel. The first coaches developed in Hungary, were uncomfortable and had no protection from the elements. Horses were at first selected for the task of pulling this new form of transport, but later were bred specially as they required both speed and stamina to fulfil this new role.

During the Tudor and Stuart period, the improvement and interest in cereals and other crops overshadowed that of farm livestock. Nonetheless the stockmen of the time ceased to be content with either the traditional methods of husbandry or the types of the livestock. The latter had improved very little overall since Roman times.

Some districts, as we have seen, developed more specialised strains of cattle, sheep and horses, but pigs had been neglected to a great extent and were still very like their Iron Age ancestors, except they were larger. It was to be a few more decades before the real strides were taken in their improvement.

With the new trade, both national and international, and the development of new markets as well as more mouths to feed, the aim of the Tudor agriculturalist was to improve his crops, livestock and methods of husbandry. The motive of profit became all-important.

Wool remained eminently saleable. Red meat, butter and cheese rose to equal importance with the golden fleece. The cow was beginning, more than before, to be looked on as a provider of meat and milk, but in some areas still retained its importance as a breeder of oxen for plough work.

The carcases and prolificacy of sheep assumed almost equal importance to its wool-producing capability. The horse needed to be improved for its new and diverse role to provide a riding animal of varying size, and an animal capable of pulling loads on the land, or coaches on the highway.

Importation of Livestock

From the time of the first farmers, who brought their livestock to Britain from Europe, importation of animals had certainly taken place. Little is recorded and the scale of importation would have been such that it only had an effect on the animals in the area where the immigrants, either hostile or friendly, settled. The Romans left their improved strains, so did the Vikings and the Saxons, and these animals which were inter-bred with indigenous or earlier imported livestock, all had a part to play in the make-up of our present-day farm animals.

With the vastly improved trade and the need for a more efficient agriculture, the Tudor and Stuart farmers took advantage of these opportunities. Writers of the time describe the new pied cattle which were beginning to appear on the scene. The milk cow of Lincolnshire was said to be taking on an increasingly important role. It was described as a lean, high-thighed animal, with strong hooves and short horns. References were made to it being of the Dutch breed and this seems to be acceptable, as much trade has always been carried out in livestock between Holland and Lincolnshire.

At that time cattle of the Low Countries and Northern Holland were of a very mixed type, a fact which is born out by the paintings of the Dutch masters such as Albert Cupt. These show cattle of many colours, but they all were said to have two characteristics in common, high yield of milk upon which the large butter and cheese export trade of Holland was built, and their considerable size. They were white and grey in colour, as well as yellow and dun. The latter type may have been the ancestor of the Suffolk Dun which later was used to cross-breed with the Red Norfolk to produce the Red Poll cattle of south-east England.

Some evidence seems to support the theory of importation of large numbers of Dutch cattle into Britain in the later sixteenth and seventeenth centuries. They were at first concentrated in Lincolnshire and Kent, but gradually spread northwards into Yorkshire and Durham where the imported cattle form the basis of the Longhorn and, later, the Shorthorn breeds. The main characteristic of these big-framed Dutch cattle was their capacity to produce a large yield of milk. The great variation in colour and shape, as shown in the old paintings, gives us a lead to the origin of some of today's minority breeds. The creamy, dun animals with mealy noses have a distinct look of the modern Jersey cow although the latter are much smaller in stature.

Other Channel Island breeds, such as the Guernsey (the now extinct Alderney cow) also appear to have taken many of their characteristics from the Dutch cattle. This whole theory stands up to more scrutiny, as during the reign of Mary there is sure to have been trade between England and the Netherlands after she married Philip, heir to the throne of Spain and the Netherlands.

Claims are made by Dutch agricultural historians, that the Longhorn and even the White-faced Hereford breed which had become well established by the end of the eighteenth century, were based on the Netherland cattle being crossed with Devon, Welsh and Lancashire stock. This seems likely as both White-faced and Longhorn animals are represented in the Dutch paintings. It is also recorded that land-owning families from Hereford procured cattle from Holland and the Low Countries.

The animals which were imported from Holland at that time, and were to improve and help in the make-up of most of our British breeds, not only the Longhorn, Shorthorn and Hereford but also the Gloucester and the Ayrshire, were unlike the modern black and white Dutch cattle, or Friesians. The latter were, to a great extent, influenced by the importation of German cattle into Holland after their own stocks were depleted by cattle plague in the eighteenth century, but both the Friesian and the Shorthorn breeds must have had ancestors in common.

In the seventeenth century, the Low Countries also supplied stocks of cattle to America, but historians record that these cattle were not as hardy as the animals from Britain. They were too specialised and unable to withstand the conditions. The medieval management in North America was similar to that of Lincolnshire and not to the liking of these more selectively-bred animals. It was said that the English milk cow was much less trouble than the cows from Holland and only required a little hay, whereas the latter required much more attention and better feeding.

Original stocking of New England took place with the Devon breed, probably taken by the Pilgrim Fathers. One Dutch historian recorded his observations that Devon cattle do not grow as large as Dutch, nor give as much milk, but are cheaper to keep and they fat and tallow well.

Cattle

Cattle in Cheshire, as described by the historian William Smith, were almost certainly the ancestors of the Longhorn. The vast amounts of milk which they gave due to their Dutch blood, was made into butter and cheese and was sent by sea from the

A Welsh Black single-suckling herd on a bleak hillside near Llyn Celyn, North Wales.

Dee and Mersey ports to London, and by road to the great dairy market of Uttoxeter. Smith described the Cheshire cattle as black, smooth-coated, with exceedingly long white horns, with black tips, very like the present-day Welsh Black breed. They were probably brought from Wales in earlier days and later crossed with the new Dutch cattle.

The Black cattle of Wales were, at this time, beginning to take on a very important agrarian role. The rise of store cattle production and the droving practises seem to suggest that store cattle were driven from Wales and Scotland for fattening on the better, more fertile pastures in England. White cattle, with red or black points, which have been mentioned previously, do not seem to have increased in importance in Wales, although some domestic herds were still in being.

On the lower land, both red and brown cattle are recorded, but the ancestors of the modern Welsh Black were in the majority, being able to thrive on land of varying quality. The very first Black cattle, as we have seen previously, had been brought millennia before by immigrants from the Iberian peninsula and France who settled on both sides of the Irish Sea, in Ireland and, it seems likely, on the western coasts of England and Wales. The ancestry of the Welsh cattle may have much in common with the fighting bulls of Spain. Until at least the beginning of the seventeenth century, some wild or feral cattle survived in Wales, as reference to the hunting of wild bulls in the Welsh hills is recorded. Whether these were white or black cattle is not stated.

In Scotland the pattern was much the same as in Wales. A few wild or feral cattle were left to roam the great forests and glens. The Kyloe, or Highland cattle, of Argyle and Ross were, to a great extent, semi-wild and were direct descendants of the pre-Roman cattle, whereas the cattle of the Western Isles were closely related to the Black cattle of Wales and Ireland, very much like the Dexter, Kerry and the Welsh Black of today.

It seems clear that during the sixteenth and seventeenth centuries the predominant

A fine Highland (Kyloe) steer owned by the Marquis of Cholmondely.

cattle of Britain were of the Black Celtic type which had long upswept horns and a very hardy disposition. They were of common ancestry and had been brought to Britain by the settlers from France and the Iberian peninsula centuries before. Their concentration along the west of the British Isles was brought about by the movement westward of the Britons and their livestock, to avoid the ravages of attacks by both the Vikings and the Saxons on the east of the country. Natural selection also favoured black cattle, with their ability to withstand the more extreme weather conditions. Through the south, east and midlands of the country, imported Dutch blood was beginning to influence the type of cattle more and more. Red and brown, whole coloured animals, with short horns, these large-framed cattle produced more milk than the black animals of the west of the country, but were unable to tolerate the poorer pastures or the damper climate.

The ancestors of the Longhorn, the Shorthorn and the Red Poll, these animals gradually spread over much of England. In the broad belt running through the centre of England to the Welsh Border, a split-coloured animal containing white markings, which was possibly a mixture of all the previously imported blood and the blood of the original and ancient White cattle, together with that of the black Celtic animals, existed. Whitehead suggests in his book *The Ancient White Cattle of Britain* that many local strains received blood, and therefore type, from the White cattle of the area, for instance the Chartley Herd influenced the Staffordshire Longhorn cattle. They are all sure to have been selected for their improved performance, as well as their hardiness and their ability to thrive.

Whalley Abbey in Lancashire, which was originally part of the Forest of Bowland, is said to have had a herd of White Polled cattle until the end of the seventeenth century and the Reverend John Storer in his book *The Wild White Cattle of Britain* states that these cattle had been obtained from the Bowland Forest at the time of the Lord Abbots. The cattle, which may have been enclosed in the park during Norman

An early pedigree Hereford bull.

An early Suffolk ox.

times, were later sent to Gisburne Park in Yorkshire and Middleton Hall in Lancashire, where they were said to have become completely tame. These animals are said by some authorities to have provided the foundation stock for the British White Breed, which is at the present time experiencing a revival of interest, due to its new-found popularity in North America and Australia.

Other local types or strains were developed at this time and the Gloucester breed, which was selected for its milking qualities and the White-faced Hereford, which became one of our foremost beef breeds, are two examples of this selection. In East Anglia a dun-coloured high milking strain was prominent and this animal is thought to be an early example of the Suffolk Dun breed; and the Red cattle of Devon which were exported to New England were again the foundation stock of important new breeds.

In remote parts of Scotland and the Western Isles, the Kyloe or Highland breed retained a foothold, as they have succeeded in doing to the present time.

Sheep

During the Tudor period the Midland Counties of England were heavily stocked with sheep, an inheritance from the monasteries after their dissolution. These vast tracts of land were now in the hands of laymen and flocks of up to 1,000 animals were not uncommon. At the end of the reign of Queen Elizabeth, mixed farms consisted of up to 10 milk cows, a few sheep, the odd pig, some oxen and poultry etc and also three or four horses.

The sheep of Leicester, as described by writers of the times, were large-boned, of a good shape with long coarse wool, more so than those to be found in the Cotswolds, and became known as Midland Longwools, as they stocked most of the

Twin Ryeland lambs at Graves Park, Sheffield.

A Portland ewe and lamb. This breed is said in folklore to have swum ashore from ships wrecked in the Spanish Armada.

Midland Counties of England. They were undoubtedly of the same type as the original Roman Longwool sheep which had found their home in the Cotswolds, Kent and Lincolnshire.

At the beginning of the seventeenth century, the southern and midland counties had no indigenous breeds of sheep of their own and were fully stocked with animals from other counties, Wales and the West Country. The lighter heath land was stocked with short-woolled breeds which were often horned, the most famous of which was the so-called Ryeland, which as we have seen, was originally bred in Herefordshire and Worcestershire. The Portland sheep, which is now a very rare breed and has only recently been saved from extinction by the Rare Breeds Survival Trust, is said to be a descendant of the unimproved south-western horned type and was common in Devon, Somerset and Dorset. The story of how the ancestors of this breed swam ashore from sunken ships of the Spanish Armada is possibly based on fact, as they have naturally long tails unlike many of our other breeds; like the Mediterranean sheep, some are able to breed all the year round. This little horned, tan-faced sheep, has passed its characteristics on to some of our commercial breeds such as the Dorset Horn.

John Leigh of Steperly Park, Adlington, Cheshire, is reported at that time to have kept a flock of multi-horned sheep. According to William Smith, the topographer, some of these animals had horns of great size and fleeces which contained a great deal of hair. Similar sheep were also reported in Staffordshire and whether these animals are the residue of an ancient breed, or exotics kept for ornamental purposes, is difficult to decide as there is no information given as to their colour. They could possibly have been an early flock of the very popular, present-day Jacob sheep (see p. 138).

A four-horned Jacob ewe with twin lambs. The breed was once extremely rare but it is now no longer endangered.

Welsh sheep during the Tudor and Stuart period had a poor reputation for wool, but one historian wrote 'Welsh sheep are all of the worst, for they are both little and of worst staple, and indeed are praised only in the dish for they are of the sweeter mutton', much as today the Welsh lamb cannot be bettered for its table appeal. The criticism of the wool was compounded by the fact that wool merchants from England dumped their worst wool in Wales, for weaving into coarse Welsh cloth.

Unimproved flocks of the same Welsh sheep have, until recent times, survived in remote pockets in the Principality. One such is the Rhiw breed, which until 1980 survived on the mountain pastures of the hill after which it was named, on the Lleyn peninsula (see p. 60).

Scottish sheep were all derived from English breeds, except on the islands where small nuclei of the old unimproved Soay, St Kilda (now called the Hebridean) and Shetland breeds still survived just as they do today (see p. 125).

The dun-faced, horned Heathland sheep of England added to the make-up of the Scottish Blackface and the Cheviot breeds. The North Ronaldsay sheep were, like sheep and cattle on other islands, starting to develop a taste for seaweed which they ate during the winter period.

A North Ronaldsay lamb (seaweed-eating sheep).

Sheep of the Stuart period fell into four groups, the small horned sheep with poor fleeces which contained a high proportion of hair or kemp and were found in Scotland and the Western Isles, and were little changed from the prehistoric sheep of Britain. The present-day Soay is a direct descendant and very similar to these animals. The Heathland type of sheep which is thought to contain the same

47

A North Ronaldsay ewe with different coloured lambs.

A Soay ewe suckling its lamb. The short (or deer) tail is a characteristic of all primitive breeds.

primitive blood, but had been selected and improved by the flock masters, was a heavier animal with coarser wool and was prevalent in the south, south-west and midlands of England and Wales.

The third group was the black-faced, hairy-fleeced, horned sheep of the Linton variety found in the north of England, from which the Scotch Black-face and the Welsh Mountain are the best known modern representatives. The horn shape seems to suggest Soay and Argali blood, so this group, too, is derived from the primitive sheep of Britain being crossed with the medieval imports of Merino or Mediterranean sheep.

The other main group is the exotic long-woolled sheep, such as the Cotswold, Lincoln, Leicester and the early Wiltshire Horn. These are all thought to be direct descendants of the old Roman stock.

Horses

The horse of the Tudor and Stuart era had been changed dramatically by land-owners, soldiers and sportsmen. The great horse which had developed for use on the battlefield came into use on the farm and bore little or no resemblance to the little-changed descendants of the prehistoric ponies which worked on the land in the north and the west of the country. Due to its strength it became popular as a replacement for a number of oxen in the working team. Though the real importance of the horse at this time as an agricultural animal was mainly due to the fact that it was bred on farms, the work to which it was put in agriculture was only of secondary importance, as oxen were still used to a great extent for heavy duties.

Horses for military and road transport purposes had a most important role to play and had been developed first from the Roman and later from the Norman importa-tions. James II is said to have brought mares from Hungary to improve his studs and later importations from Spain and France were used to develop the much valued war-horses of the period. Further imports from Poland, Denmark and Sweden, provided new blood with which to increase the size of the heavy working horse.

The measures taken during this period to restrain the breeding from undesirable stock had a profound effect on the type of animals which were selected, and at the end of this period the purpose-bred horse had begun to take its place.

Pack horses and animals to draw the new, roofed, leather-covered coaches were now required in greater numbers. Different requirements were needed for the varying jobs of work and depended to what use the horse was put. By the end of the seventeenth century the newly developed animals could produce speed, endurance and strength.

Pigs

Pigs of the Tudor and Stuart period had changed very little from their early ances-tors, but in certain areas of the country due to different forms of management and better feeding techniques as well as more rigorous selection, local strains had de-veloped even further. Some imports may have taken place, but whether this is true or not it is difficult to decide. In Leicestershire and Staffordshire, where the new management had been practised the old type Tamworth pig may have appeared on

Ten-week old Tamworth piglets. This is Britain's only red breed of pigs.

the scene. Gervase Markham, writing in the seventeenth century, said when discussing pigs, 'the colour is best which is all in one place, as all white or all sanded, the pied are the worst and most apt to take measles, the black is tolerable, but our kingdom through coldness breedeth them seldom'. The sanded pig may have been the early Berkshire or Tamworth, and the white may be early examples of pigs from Wales or Cornwall.

The 1600s left agriculture with many new ideas and an abundance of widely differing types of farm animals which were to be used by livestock improvers of later centuries.

5 The Eighteenth Century

George III was the first of the Hanoverian line to be born in England. It was the king's political opponents who first gave him the nickname 'Farmer George'. By giving him such a name and trying to make him out to be a figure of fun and ridicule they did, in fact, endear him to his subjects as well as future generations. In reality he had a deep love for and intense interest in farming. He corresponded with agricultural writers of the day and it was reported that he wrote under an assumed name in agricultural journals.

During his reign great strides were taken in the breeding of livestock. A far more scientific approach had started in place of the haphazard method of previous generations, and the king made an important contribution to agriculture, not only in Britain but further afield. He inherited the Richmond Park estate from his grandfather George II, in 1760. It had been laid out in the new natural style earlier in the century by Queen Caroline and was one of the first estates to have been changed from the rather formal style of William III's time.

The king exercised his passion for farming and employed the best-known planner of the time Lancelot, or 'Capability' as he was known, Brown, to convert the estate for farming use. In 1772 he inherited the remainder of the estate from his mother, Princess Augusta, nine acres of which included the newly established Botanical Gardens. Much of the remainder he stocked with both cattle and sheep and took

An example of an early imported Merino sheep from Spain.

under his direct control the 300 acre estate where he could continue to enjoy his farming pastime.

At this time in the eighteenth century, English wool was very coarse, due to it containing a high proportion of the waterproofing fibre or kemp. This made the wool unsuitable for the production of fine cloth, and the practice for many years had been to import finer wool from Spain which was the product of the Merino sheep, to mix with the home-produced article. The imported wool, due to it being a product of animals of a drier climate, contained very little of the kemp and was, therefore, of high quality. This fine wool was the subject of a vast trade between Spain and Britain. It is said that one day the king and his advisor, Joseph Banks, discussed the possibility of importing some of the merino sheep to Britain to improve the quality of the wool produced by the royal flock, which at that time was made up almost exclusively of Wiltshire Horn sheep.

It proved difficult for some time to obtain the sheep from Spain, although some flocks had been established in other parts of Europe. Eventually, on 4th April 1788, the first three merino sheep arrived at Kew to be followed later by more importations. By the year 1793, after many troubles and disappointments, the king's merino flock numbered about 150 animals. To prevent more disasters the sheep were moved to Windsor, another royal estate, where it was thought they would thrive more easily, and during the next ten years the flock number remained at about 100 animals, so it was decided to return them to their former home at Kew. Sir Joseph Banks advised the king that the time had come to hold a sale of surplus animals. The original plan had been to spread the breed over the country, once it had become established at Kew. On 15th August 1804, fifty people gathered under the trees near the pagoda in the field which is now part of Kew Gardens, for the first Royal Sheep Sale.

The Agricultural Revolution

During the hundred years when England was ruled by the three Georges, changes in agricultural methods took place to revolutionise the whole industry. Livestock took on a still more important role as the population increased still further and resulted in more and more mouths to feed. The livestock farmers of the eighteenth century inherited regional types of stock which had been modified over the previous century by the introduction of Dutch blood into cattle, some merino blood into sheep, and miscellaneous European strains into the horse. They also inherited the new techniques of management and new fodder crops, and with these a new spirit of scientific and theoretical enquiry.

Fast-growing demand for livestock products, meat, milk, cheese and good quality wool, could only be met by the British farmer and for the following 250 years he tried to meet the demands put on him for the products from the home market. Cattle, sheep, pigs and horses came to the fore and the triumph of 'horn over corn' took shape in the farmlands of Britain.

By this time, the supply line of meat and other livestock products, especially to the London market but also to other centres, had extended to the remote highlands of Scotland and the Welsh mountains and also the extreme south-west peninsula of Cornwall. The majority of the store cattle at the beginning of the century, which

An old-fashioned cow house. Note the wooden pails and milking stool.

came from the west of Britain, were still the longhorn, Black Celtic cattle, but other colours were beginning to appear more and more. It was not until a board of agriculture surveyors analysed the regional variation in cattle at the beginning of the nineteenth century that a clear picture of the type of cattle at that time can be formulated. The Welsh stores or runts, along with the Scottish cattle, were still brought and fattened on the better pastures of southern England and it was a common sight to see these black cattle, and the bigger, coloured cattle from different parts of England, all being grazed on the fattening pastures just a few droving days from London, where they were destined for the Smithfield market.

With the introduction of the railways and the steamships, fattening was carried out on the more remote pastures of Wales, Cheshire and Herefordshire, within a few hours' or days' travelling time from their eventual destination. Around London the good grasslands, as well as the farms which produced agricultural by-products, were in constant use for finishing both the cattle and sheep for the meat market. Many of the fields which were used during that period still retain a vast amount of fertility due to the droppings of the fattening animals. Romney Marsh was a very important fattening area, much as it is today.

Cattle

Milk was taking on a role of increasing importance. Due to its being perishable it did not, at this time, join the influx of food into London. It was used, to a small extent, to make puddings and in tea, and the fashion became more popular when the first city cow-keepers appeared on the scene. They paid high rents for suburban pastures to produce fresh milk to supply the towns, but by and large the majority of the milk which was produced in the countryside ended up as salted butter or cheese, as these products were so much more easily transported and had a much longer life than raw milk.

In the early part of the century, there was still very little difference between the milking and the meat-producing stock, although at this time some types were beginning to be reported 'better for the pail', or more suitable for the butcher, but even the Black cattle of the highlands were expected to milk into the churn as well as breed a store for the drover.

Processed milk flooded into London from all parts of the British Isles and it is recorded that Cheshire sent 14,000 tons of cheese every year to the capital.

By the middle of the eighteenth century, the type of dairy cattle used to produce milk for cheese-making varied considerably from area to area and according to

A Longhorn cow. Note the white back stripe which is carried by many ancient breeds, being originally a form of camouflage for their wild ancestors.

Robert Trow-Smith in his *History of British Livestock Husbandry* even individual herds of cattle varied in shape, size and colour, being black, red, brindle and often having a white stripe along the back which carried on down the tail. This is said to be inherited from the original wild cattle and was a characteristic intended to help camouflage them by breaking up their shape in their natural woodland home. The type so described at this time appears to have been a conglomerate of the original Black longhorned Celtic cattle, the Yorkshire Shorthorn as it was called, and the Midland Longhorn, and, although mixed herds were usual, the early livestock improvers were beginning to make their mark on the newly emerging breeds.

In Gloucestershire a type of dairy cow, that was said to be responsible for the popularity of Double Gloucester cheese due to the small fat globules of its milk, was

A Longhorn cow with downswept horns.

beginning to become popular. The Gloucester breed, which is said to have been a strain which was originally bred in Glamorgan and is now very rare, was saved from extinction by the Rare Breeds Survival Trust and has changed very little in colour to this day, being dark mahogany with the primitive white stripe extending from the middle of its back over its tail, down its hindquarters and forward under its stomach to its front legs. It is said to have been developed from the cattle of Wales with a dash of Dutch blood. To what extent the introduction of Dutch blood affected cattle is difficult to say, but the Durham or north country oxen, whose mature weights were the wonder of the age, certainly benefited from this introduction.

Breeds developed more quickly as the century progressed but the foundation stock of all the cattle were the primitive Black longhorned cattle of the north and west, with the added Dutch blood to a lesser or greater degree. The old breed provided the thrift and hardiness, whilst the new importation improved the performance. Although polled, or hornless, strains had appeared from time to time in all cattle, it was not until this period, apart from the early selection probably made by the Vikings, that this characteristic was selected for. The first steps took place in northern Britain where the carriers of this hornless characteristic were deliberately

An Aberdeen Angus single-suckler cow with her own cross-bred Charolais calf.

Part of a herd of Belted Galloways grazing on the Kirkcudbrightshire hills.

multiplied. The polled gene is dominant and easily bred for, as one parent carrying this character invariably throws a polled offspring. Rapid multiplication and selection can take place, and at this time the two famous Scottish breeds of the Aberdeen Angus and the Galloway were emerging independently and along different lines.

For practical reasons, another characteristic which was selected and bred for was the shorter horn type of cattle, known as the Shorthorn. As the density of cattle increased, long horns as a physical characteristic began to go out of favour.

At the beginning of the eighteenth century, all cattle were triple purpose, used for beef, milk and traction. The Black cow was adequate for the needs of the still relatively primitive agriculture in the north-west of the country, but as the century progressed different types emerged. In the lowlands, south-east of the Wash, the West Midlands and the Bristol Channel southwards, a reddish or brown middle-horn type of cattle had begun to develop. These were ancestors of the Devon and Sussex cattle and were still multi-purpose, but did give a higher milk yield than the old Black cows.

An engraving of an early Shorthorn bull.

Much of the white colour which was beginning to appear may have been the result of crossing with the indigenous herds of White cattle.

By the middle of the eighteenth century, three types of cattle begin to emerge, the Black Longhorn, the Red or Brown Middlehorn and the Southern Counties Broken-colour. The Hereford type of cattle developed from the injection of Dutch blood into the red cattle of the Midlands. The two types of Shorthorn, the Dairy (or Yorkshire) and the Scottish (or Beef Shorthorn) had begun to arrive on the scene.

At this time only the Longhorn and the Shorthorn were of national importance, while all other types remained within their region. The Red Devon cattle had spread to a certain extent from their native area, due to their good beef and milking qualities. Longhorn improvement had been progressing for some time because in 1810 a Mr Webster was said to have brought some of these cattle from the banks of the River Trent to Canley, near Coventry, in Warwickshire. These animals were said to have been developed east of Morecambe Bay and were the direct result of the crossing of the Celtic Black Longhorn cattle with Dutch stock. The latter, it was said, gave the animals both substance and white markings. From there they spread southwards into Lancashire and became known as Lancashire Longhorns. From these cattle Sir Thomas Gressley selected animals with which to build his herd, and from this herd Mr Webster bought his cattle, which were said at the time to have been red animals containing only very little Dutch blood. Webster selected his first stock very carefully and he became the first livestock breeder to name an animal, thus identifying it in the pedigrees of his improved cattle. This bull was called Bloxidge, and was of pure Canley blood and featured in the early work of most of the pioneer breeders. Bakewell used this blood for all his breeding experiments.

The call for more meat gave the first opportunities to the real livestock improvers. The Georgians were big meat eaters and this made up a major part of their diet. As the population grew in the industrial areas as well as in the big cities, so the need for better stock to produce yet more meat became important. Robert Bakewell gave Britain the name 'the Livestock Market of the World' and he is said to have produced the first 'roast beefe of olde Englande'. Up to Bakewell's time the Longhorn had been a dual-purpose animal and every county appeared to have its own Longhorn type, from the semi-wild Longhorns of the west coast to the more kindly Longhorns of the Cheshire Plain.

Robert Bakewell is the best known of the livestock improvers, although he had made use of cattle selected by other people such as Webster, and before him Sir Thomas Gressley. He was one of the first people to use a planned breeding policy for livestock. He was a born showman and this, no doubt, helped to promote his

An engraving which illustrates Bakewell's New Longhorn cattle and New Leicester sheep.

name during his lifetime; even today he is still the best remembered of the series of livestock improvers. He was never communicative about his breeding methods, but was always ready to display the excellence of his stock. He was born in 1727 at Dishley Grange, near Loughborough, in Leicestershire, and succeeded to the management of his father's farm in 1760, the year of George III's succession to the throne. He had been experimenting with animal breeding for the previous fifteen years and by the time he took over the family business he had already formulated the policy he intended to carry out in the improvement of livestock. He not only preached but also practised his art and it became the spirit of the age. His claim to fame was not only in his breeding successes, but also in his idea of hiring out sires to evaluate their worth more quickly. He was the first livestock breeder to concentrate solely on pedigree breeding techniques and he originated the first progeny testing.

Through his arrogance, showmanship and new ideas, he was not always very popular. As is often the case with people who set their face against custom, he met with many difficulties, not the least when he became bankrupt in 1776.

Prior to Bakewell's improvements, the majority of Longhorns were coarse-boned animals whose colour ranged from red, pied and brindle, to black, and often they had the finch back of the primitive cattle. But they did produce a reasonable quantity of milk as was seen from the amount of cheese produced in Cheshire. The bullocks had always been expected to fatten for meat, in fact the breed provided dual-purpose animals. Bakewell decided to develop the fattening potential and selected the animals which he thought had the qualities which he required. With the Canley stock which he obtained for his work he used close-breeding techniques and soon began to produce the much improved Longhorn, but in the process greatly reduced both the breeding and milking capabilities. This, again, did not please many of his friends, for instance those who made the famous cheese and needed an animal which could produce a large quantity of milk of high quality.

His two most famous bulls were 'Twopenny' and 'Shakespeare', and they were to be found in the pedigrees of most of his animals. They were respectively a son and a grandson of Webster's Canley cows.

As the name suggests, the Longhorn had a pair of very long picturesque horns which were a most inconvenient characteristic. This was the main reason why the breed fell out of favour and was overtaken by another breed, the Shorthorn. During the last few years, Longhorns have made somewhat of a comeback partially due to the fact that cattle can be disbudded at a few days old. This process involves killing the horn buds, preventing the horns from growing. The breed is also used for single suckling systems of producing beef and many stately homes, too, now have a herd of Longhorn cattle to enhance their parkland as well as to produce a good financial return.

The qualities which Bakewell was said to require in his cattle were small bones and the most meat at the most valuable joints. He said that the smaller the bones the quicker the animal will fatten. Another point which he classed as important was the power of a good beef animal to pass on its superior beef characteristics to its progeny.

Bakewell's pupil, Charles Colling, used many of the principles of breeding which he had learned from his teacher and developed the newer type of cattle, the Short-horn, which was then coming into predominance. Previously this type had been known as the Durham, or North Yorkshire. The prototype is said to have been a cross between the indigenous red and black cattle and Dutch stock. It probably contained some Highland and Galloway blood and was found in the main in the North Yorkshire area. Its original home had been in Lincolnshire, for in earlier writings it was referred to as the Dutch breed of Lincoln, and was known to be a most profitable beast for the dairyman, grazier and butcher. The colour was black or red, but the common characteristic of this type was their short, curved horns. The more efficient qualities of the pure Dutch cattle, such as higher milk production of better quality and better fattening potential, was also accompanied by a loss of hardiness, and they were criticised for being unthrifty and unprofitable under hard conditions. Once crossed with British stock this fault was rectified and the thrift of the old Black cattle, plus the bulk and the yield of imported animals, combined to produce the Shorthorn breed.

Local Shorthorn strains developed over the years, of which the Durham and Teeswater were the most famous. Charles Colling and his brother developed the Durham to such an extent that in their cattle sale of 1810 twenty heifers produced an averge price of £140 4s. 7d. (£140.23) and ten bulls £168. 8s. 0d. (£168.40), but the most outstanding amount of all, 1,000 guineas, was paid for a bull named 'Comet 155'. These prices at today's values would be astronomical, but they just prove how admired the animals of the livestock improvers were in their day.

The Colling brothers used two of their most famous animals, the Durham Ox and the White Heifer, both of which were sired by a bull called 'Favourite 252', in an intensive campaign to publicise their new improved Shorthorn cattle, and the breed remained the most numerous one in the country until the middle of the present century. It was developed later into three different strains, the beef Shorthorn which was concentrated in Scotland, the dairy Shorthorn in the Yorkshire area, and the latter was sub-divided again into the Lincoln Shorthorn.

Sheep

Around 1700, the sheep of Britain were still of the four basic types described on pp. 44–9. Welsh sheep flocks grew vastly due to the home and factory weaving and the great need for wool. The Welsh breed was crossed with English rams to improve their type and the first improved Welsh breed was the Lleyn, a distinct variety which is still confined to its original home in Carnarvonshire and on Anglesey. The first Lleyn sheep were said to have been produced by Mr Lloyd Evans of Nan Horen, who crossed his ewes with a Border Leicester ram in about 1810 (although Roscommon rams may also have been used). Some experts say the original Welsh sheep was of the Roscommon type, which was later represented in Wales by the Rhiw breed which was found on the tip of the Lleyn peninsula by Rhiw Hill. This breed is now thought to be extinct, but it is said that selections from it eventually evolved to produce the Lleyn breed.

Professor J.B. Owen of Bangor says, 'Two types of sheep have been associated with North Wales for a very long time, the small Welsh Mountain and the large polled breed with the white face and legs. The latter seems to have a firm association with the Lleyn Peninsula and the development of the Lleyn breed. Renowned for its prolificacy and its mothering ability, the breed was widespread in the lowlands of the northwest corner of Wales at the end of the nineteenth century and up to the 1930s. During the middle of the century the number of Lleyn sheep declined and the breed became in danger of extinction, due mainly to the availability of cheap, draft, hill ewes from the Snowdonian flocks.'

In the late 1960s the breed reached a turning point and has since flourished. Cymdeithas Defiad Lleyn (Lleyn Sheep Society) was formed in October 1970. The renaissance of the breed lies in the character of the animals themselves. High

Welsh Mountain sheep. Note the ram's hairy mane and beard. The wool contains a high proportion of hair and is typical of a mountain breed.

prolificacy, attractive body conformation, together with their excellent mothering qualities, have found favour in many parts of Britain where they produce ideal lambs, both purebred or when crossed with Down or Texel rams.

At this time the Cheviot breed was emerging in Scotland, being again a cross between the Leicester sheep and the indigenous breeds.

Lleyn sheep at Tan-y-Bryn, Sarn Bach, Abersoch. The ram is wearing a market harness.

Most sheep in this period contained some horned individuals, in fact hornlessness during the early eighteenth century was rare, and complete flocks which did not have horns were referred to as 'pole' sheep. British sheep were hardy, unlike the continental breeds, and were rarely housed in the winter-time. A writer of the time wrote 'British sheep go constantly under the open sky'.

The new fodder crops being developed by people such as Townsend boosted the development of improved breeds which needed to be less dependent on their ability to forage for a living or thrive on poor grazing.

At this time, Bakewell was experimenting with his new Leicester breed. The longhorn fever which he had instigated had been mainly localised in the Midland Counties but the epidemic of fashion in the Bakewell New Leicester was nationwide. Even his critics who did not acknowledge his work with Longhorn cattle praised

A Teeswater ewe and lamb. The breed was originated in Teesdale, Co. Durham.

what he was able to achieve with the old Leicester sheep in converting it into the new butcher's beast. Some writers of note, including Arthur Young, a respected agricultural writer of the time, said that Bakewell had used the Lincoln breed to attain this improvement.

The first breeder of note to use the Leicester sheep was Joseph Allom of Clifton. The old, unimproved Leicester was one of the family of long-woolled sheep which had been famous for centuries and had been derived originally from the Lincoln Longwool, along with the Cotswold and Romney Marsh (or Kent), and were a legacy of the Roman sheep farmers.

The Teeswater was also a member of this family, being related to the Lincoln, but was heavier, lanker and with finer bones. These longwools are more correctly described as middle-wools, the difference in the fleece being brought about by the quality of the feed. They were all hornless and white-legged, but regional difference did occur due to breeder's fancy and also the type of land on which they were kept.

Bakewell is said to have chosen his local breed to use as his foundation stock, but he possibly used either the improved Leicesters bred by Joseph Allom of Clifton or the Lincoln breed. In fact he may have used a number of different breeds. He travelled round the neighbouring farms, selecting animals which he thought had perfect symmetry and showed an aptitude to fatten. Arthur Young said that he had crossed a Lincoln sheep with a Yorkshire Wold ewe. Others said it was just an improved strain of the Lincoln, but this Bakewell never divulged. It was even suggested that he used the very old Ryeland breed. Whichever breed he used, the

A Border Leicester ewe, a direct descendant of Bakewell's Improved Leicester sheep.

animals which he bought were usually small and compact, smaller than the sheep which were then in general use and, by crossing these animals and close-breeding (i.e. mating closely-related animals together) he developed a new strain. They became known as the 'barrel on four short legs' and were said to produce good mutton one year earlier than any other breed. Again, he used his showmanship to the full by retaining a great air of mystery in all his dealings, but was nevertheless very successful. His main income was derived from hiring out his rams. This enabled him to be more selective in his breeding policy, as he had more progeny to select from. The first rams he hired out in 1760 at sixteen shillings (80p) each for the season, but at the height of his success in the mid 1780s he charged two thousand guineas for the

A pedigree Ryeland ram owned by Sheffield Metropolitan District Council.

hire of seven rams. In today's terms this is a vast amount of money and would not be paralleled except perhaps in the business of racehorse breeding. It was he who set the fashion that the dearest is always the best. Today the breed has developed into the Border Leicester sheep, and has been the foundation sire for the Welsh Mountain and the Cheviot breeds which are now used by commercial sheep farmers for the production of the Half-bred.

This was a time of great change in the sheep industry and many other breeds were beginning to take shape, but most were the result of other strains being crossed with Bakewell's new improved Leicester rams. The Cotswold, too, had taken on a great deal of the new Leicester characteristic, due to the latter's use on the Gloucestershire flocks.

Some local strains were said to have been spoiled by the new fashion of using

Dishley blood. For instance, neither the Ryeland nor the Dorset Horn breeds benefited from this crossing, but the new Cheviot which was originally one of the tan-faced breeds developed into a much improved animal with its now familiar white face. The new Leicester blood had improved it both in conformation and carcase quality. Without doubt, the addition of new Leicester blood to most other breeds resulted in an increase in size and early maturity which the meat market needed, without detriment to the constitution or merits of the breed upon which it was sparingly used, especially by some breeders who used Dishley animals with great skill. The modern Cheviot is an example of such a breeding programme.

Bakewell was the catalyst who inspired others to use his animals as a basis for improving their own flocks and herds and they were encouraged to continue the work of further selection.

Pigs

There is little information to be found about the pigs of this period. Regional types are certain to have been in the making and reference is made to the importation of the Berkshire breed into Ireland where it had a record of high prolificacy.

The Welsh pig was valued in Wiltshire and Gloucestershire for feeding on the whey, a by-product of the production of Gloucester Double Cheese, but whether this animal is the true ancestor of the modern Welsh pig it is difficult to say.

References in the *Practical Farmer* of 1732 to a small wild black China or West Indian pig being imported into Britain seem to have begun a revolution in the pig breeding industry, by bringing about a change both in the conformation and the performance of the native breeds. At that time, the fatter the meat was the higher the esteem in which it was held. Unlike the present time, the public did not then have advice on what they should or should not eat, and the new methods of feeding by-products to the new type of pig caused them to grow very quickly and lay down fat, the exact opposite to the slower maturing old pig which was very lean. This new practice became very popular and distillers maintained large herds of swine to fatten on spent grains and later sold them to butchers for the London trade.

Pigs were also fattened at starch factories on wheat offals (husks), and reports at the time describe how clean the pig housing was kept, being thoroughly washed every day. It was reported that up to 80,000 pigs a year were fed for the London Market. Domestic brewery grains were also used for fattening pigs, both at the manor house and the farm cottage, and many animals were reared in this way until the brewing ceased in about the middle of May each year.

Pig fattening became more and more popular as the century progressed, and in 1740 farmers in the Home Counties petitioned parliament against what they alleged to be unfair competition on the part of distillers who could fatten their pigs more cheaply on husks of the grain which was a by-product of the starch industry. The fattening of pigs on a large scale developed almost into an industrial process.

Where cheese was made, pig fattening became an even more important business. Whey was used for fattening in counties such as Wiltshire and Gloucestershire, and it was said that their bacon was an important by-product and was of the very highest quality. This was usually sent to the London markets, especially Smithfield.

Large herds of pigs were still in existence in Hampshire, where they were fed in the traditional way on the forest bounty of acorns and beechmast. In the middle of the eighteenth century large herds were also to be found in forest areas such as Hertforshire and the Chilterns. In areas where pigs were grazed at pasture, they were usually ringed through the snout to prevent them rooting.

In 1760 Bakewell crossed the European wild pigs with the new Chinese strain, to produce a white pig which seems to have been the forerunner of the modern Large White breed. It was then known as the Yorkshire pig, due to the fact that the European pigs which he used were said to have been a selected strain from that county.

By the late eighteenth century, it seems certain that white, ginger, black, various spotted, and partly coloured pigs were to be found, and Gervase Markham's description of animals in the late 1600s seemed to indicate that at least various types had been imported into Britain by that time. Up to 1800, it appears that no true breeds of pigs had been developed, but different colours and regional types did exist.

Horses

By this time, workhorses had developed into animals of exceptional size and strength in the Home Counties and once again Robert Bakewell's name crops up as an early improver. He took the old English black carthorse and, through selection, developed it for even greater strength and size. This is said to be another step along the path to building the Shire horse breed of England.

An engraving of an early riding or pack horse.

A young Clydesdale gelding.

Suffolk work horses were also reported to be famous for their use on farms in that area of England and from these the Suffolk Punch breed is thought to have been developed. On the higher land and in the west of the country, the Shetland and Exmoor type of pony was still being used. They varied between nine and eleven hands in height, and were used for light jobs and for hack work.

The first Clydesdale horses are said to have appeared about this time and they were developed by crossing the larger type Scottish pony with stallions brought from Belgium, but they did not at this time take the place of the medium-sized ponies which were still used in the hilly and inaccessible districts.

In the last decade of the eighteenth century and the beginning of the nineteenth, Bakewell's stock, especially his new Leicesters, had spread over the face of Britain. The new Southdown had been developed on the chalk hills of Kent by John Ellman of Glynde, and was one of the most fashionable breeds of the day. The Longhorn, the Chinese pig and the old black carthorse had become established types of farm livestock. Model steadings (farm buildings, cow-sheds, dairies etc) had been built and new fodder crops had been discovered. Breeds had been developed and varied greatly, but few had really spread outside their own regional boundaries.

The Longhorn was found from central Scotland in the north to Cornwall in the south-west. The Shorthorn was spreading more and more from its centre in Lancashire and Yorkshire. The Devon, the Gloucester, and other minor breeds in Somerset, were confined to their own and immediately adjacent counties. The Hereford was concentrated in the Welsh marshes, but was becoming popular too, further afield. The Sussex was unknown outside the county. Channel Island breeds were making their mark more and more. Originally, they had been called Alderneys, but during the end of the nineteenth century they were sub-divided into Jerseys and Guernseys.

Early Ayrshires and related types were being developed in the south-west of Scotland in the late eighteenth century and spread even more in that area in the nineteenth century, gaining a good local reputation.

The Longhorn was, without doubt, the most important cow early in the era. It was the only national breed, but regional types of Longhorn were still obvious, from the Welsh Black and the Kyloe, to the Cheshire Dairy Longhorn.

The Shorthorn was making its mark as the decade went on and by the end of the eighteenth century, was to be found in as many varied types as the Longhorn. This was due to the introduction of Channel Island or Normandy blood in the early part of the century and was intended to improve both the yield and quality of milk. During the eighteenth century, writers of the time largely ignored this breed as it was thought to be not as important as the Longhorn. The Shorthorn was, until the beginning of the nineteenth century, still referred to as the Dutch cow, or Teesdale.

A typical Ayrshire, Britain's first true dairy breed.

It was said to have short horns, a good body, a thick hide, a short coat, and a lack of hardiness, but it did give a far larger milk yield.

In the eighteenth century, colour variation was to be found in the Shorthorn, but the breed was gradually becoming generally more red and white, or a mixture of the two or flecked. At the end of the eighteenth century some Shorthorn animals were still being described as being black-blotched, and these could possibly have common ancestry with the modern Friesian or Holstein cattle of today; although the modern

Friesian is said to have originated in Germany and was imported into Holland to make up the loss when their stocks of cattle were killed by a plague of rinderpest in the nineteenth century. Before that time all Dutch cattle had been yellow, grey or blue in colour. The black and white strain probably originated from that movement of livestock and came from North and Central Europe. Some of these animals may have entered Britain at that time to give some of the early Shorthorns their black and white colour.

The Holderness type of Shorthorn cow, which was common in Cleveland at the end of the eighteenth century, was said to have been brought about by the introduction of Channel Island or French blood and this is borne out by the horn shape and the tinge of yellow in the coat colour. This was said to have been the first step in the creation of the true Dairy Shorthorn.

By the nineteenth century, the Shorthorn began to divide into its different varieties, the one type being of poorer conformation and containing Channel Island blood, but also giving a high milk yield of good quality, and the other type being the greatly improved Beef Shorthorn. One of the problems with the Shorthorn, before the introduction of the Channel Island blood, had been the poor quality of its milk, thus giving less cheese or butter per gallon. Originally, the Longhorn had been better in this respect and it also had a better beef finishing quality on worse pastures. The Shorthorn was also said to have been a more difficult calver.

As the century went on, strains of Shorthorn were improved and selected and there was some mingling of both the best of the Longhorn and the Shorthorn breeds.

6 The Nineteenth Century Pure Breeds

In the first part of the nineteenth century progress in agriculture took the form of enclosure of open fields and the advent of crop rotation. Salted beef became a popular food in preference to the fresh meat which at that time was often bad. Farm labourers starved due to the high price of food, and there were many riots as a result of this. With the beginning of the Industrial Revolution, therefore, many people left the land to work on building canals, roads and tunnels.

Until the late 1800s the motive to improve farm livestock came from the meat market which was expanding as it had done over the past two centuries, but by this time the expansion had accelerated considerably.

By the last decade of the nineteenth century, milk sales had increased dramatically not only due to the new forms of transport which had become available, but also to the setting up of the large town dairies where cows were milked in the centres of the populations, thus ensuring a supply of fresh milk. This was the stage at which the real division between beef and dairy breeds took place. With the use of refrigerated ships starting in about 1870, the home-grown beef and mutton trades were gradually being affected by the competition from imported meat, but milk sales, being unaffected, escalated enormously.

Cattle

The Shorthorn type of cattle spread first to cover the north Midlands and northern England, where their breeding and pedigrees were gathered together in Coates' *Herd Book*. This publication was in some quarters regarded almost as a second Bible. As a result of the early work of the Colling brothers and later other breeders, especially Thomas Booth and his sons, and Thomas Bates, the Shorthorn type became more standardised and was to populate the pastures of many parts of the world, and until almost the middle of the twentieth century become the major breed in Britain. The Shorthorn characteristic made the animals much easier to handle than the old Longhorn type and as herds of dairy cattle increased in numbers, so the need and popularity of the Shorthorn grew. Due to the careful breeding of a few individuals, the breed became thrifty and efficient and the animal's popularity and reputation spread worldwide.

In 1780 the Collings collected good females from various sources, but their bull known as the 'Hubback' bull was both potent and of the right type, having good conformation without coarseness of bone, and it was he who stamped his characteristics on the Shorthorn breed. By mating closely-related animals together and inbreeding, the Collings multiplied this type of animal and, with good management and a certain amount of good luck, produced the animal for which they had aimed.

There is little doubt that the Hubback's pedigree contained both Dutch blood and that of the old Celtic Highland, or Kyloe, thus combining the hardiness of the British breed with the efficiency of the Dutch. To a lesser extent than Bakewell the Colling brothers were showmen, for they produced two famous cattle, the Durham Ox which weighed 27 cwt at five years old, and the 'White Heifer that Travelled', which were displayed at exhibitions around the countryside.

It was Thomas Bates who did much to restore the milk-producing qualities of the Shorthorn for, as had happened previously with the Longhorn, this facet had been

An old sketch of a herd of Shorthorn cattle by Julius Ibbotson Free, 1816.

bred out due to the concentration on the production of a beef animal. Until he founded his Kirklevington herd, it was left to the commercial, rather than the pedigree, livestock producers to retain the more dairy type of Shorthorn. The need for the working oxen had all but diminished by the end of the nineteenth century and the new, smaller type of cattle took on two roles, those of beef and milk production, instead of the three which they had previously held. This trend did not take place in Europe to the same extent and the retention of the ox-breeding strains, which were discarded in Britain, has during the present time led to an influx of these bigger, more muscular animals into this country, to cater for the new trend in lean meat production. (This is discussed in the next chapter.)

The constant need for improvement carried on and much progress was achieved by crossing one breed with another. This practice was common until the end of the nineteenth century when herd books were closed in an effort to fix certain breed types. The criticism of the improved Shorthorn, which continued throughout the nineteenth century, was its inability to fill the pail, and agricultural writers of that time were scathing in their condemnation of pedigree cattle. Some local strains retained the good milk-producing qualities, but were still of a very mixed type. Commercial livestock men, as distinct from pedigree breeders, tried to rectify this deficiency by using improved Shorthorn bulls on unimproved Shorthorns of the old

Holderness or Yorkshire type. The improved Shorthorn due to its having been carefully selected became red, roan or white, or a mixture of all three. The latter colour is thought by Whitehead to have been the result of the introduction of wild white cattle blood, while the commercial Shorthorns still had a range of colours, from red through white to brindle and black.

At the end of the nineteenth century, the Shorthorn type had finally taken three different directions: one resulted in the Scottish Beef Shorthorn which was to become for some years the premier beef animal of northern Britain; another was the North Country Shorthorn which was rarely seen outside its region, but retained the true dual roles of milk and beef production; and another offshoot of the latter, which varied only in the colour of its coat, was called the Lincoln Red Shorthorn, due to its deep cherry-red colour. It was first developed at the beginning of the nineteenth century, as a fancy by Thomas Turnhill of Beasley in Lincolnshire.

In 1833, consignments of improved Shorthorns were imported by the Ohio Importing Company into America, and by 1839 the Shorthorn had taken the place of the Longhorn in Ireland.

In Scotland the regional types of cattle which had been developing during the eighteenth century had, by the nineteenth, become the basis for new breeds. The Galloway cattle of south-west Scotland, which had developed a polled characteristic, were at the time to be found in a variety of colours, but black was the most popular, favoured particularly by the fatteners of England. It had retained its hardiness and ability to thrive on bleak pastures but, over the years, had to a great extent lost its ability to produce a reasonable milk yield. Originally, as reported by Robert Trow-Smith, the Galloway fed a calf and supplied a little milk, and he quotes the Second Board Survey of Ayr which stated, 'The (milk) maid takes the near and the calf the far side of the cow and both exert themselves to extract their share of the treasure. The struggle (for milk) between the maid for the family, and the calf, is renewed twice every day so long as the cow gives milk.'

A belted strain of Galloway was developed at the end of the nineteenth century and is now relatively rare, there being only about 700–1,000 breeding females at the present time. Galloways are famous still not only for their own intrinsic value but also for their ability, if crossed with a white Shorthorn bull, to produce the well-known blue/grey offspring. This cross first became popular in about 1860.

The Aberdeen Angus was developed from an old strain of Polled Black cattle, which may have existed in Roman times, by Hugh Watson who farmed in Angus. The breed also lost its milking capability and eventually overtook the Galloway in popularity as the premier Black beef breed of Scotland.

The Highland, or Kyloe, breed was improved by the Stewart family from Harris, and still remains an important beef animal in exposed areas where no other breed could possibly exist.

The English breeds, the Hereford, Devon and Sussex, were localised in their own regions, whereas the Shorthorn was widely distributed. The Hereford spread to the adjoining counties and, to a certain extent, into Wales, filling the vacuum left there by the Longhorn. The first volume of the *Hereford Herd Book* was published in 1846, and from this date onwards Herefords improved in conformation and their popularity increased. They also became popular abroad and were sold to ranchers in North and South America and Australia. The first American imports were said to have taken place in 1817, but it was not until about 1875 that the Hereford took on an important role on the ranches of the South and Middle West, where it replaced the old Texas Longhorn to a great extent, and after this time left many white-faced progeny, this being the hallmark which it passes on to all its offspring.

The Welsh cattle were mainly still of the primitive Black type, but were during the eighteenth century merging to form one breed, the Welsh Black. Although little selection took place and improvement was slow, the cattle did retain their inherent hardiness. One regional type of Welsh cattle, the Glamorgan, which is said to have been the forerunner of the Gloucester due to it having a similar colour and white back stripe, died out as it had no economic role to play in that area after the end of the eighteenth century.

The North Wales Welsh Blacks, then to be found only in Carnarvonshire and Anglesey, were still small and rangy, whereas the Cardigan type had improved due to the introduction of West Highland blood by the Duke of Newcastle. The red cows of Montgomery were said by Robert Trow-Smith to have been replaced by the Hereford breed from the neighbouring county.

The Castle Martin, or Pembroke cow, improved more than any other of the Welsh breeds, due it is said by some authorities to the introduction of still more English and Dutch blood. Whitehead is of the opinion that this improvement was probably brought about by the use of bulls from the herd of White cattle at Dynevor Castle, which had been domesticated years before and used on the estate for milk and beef production and for draught purposes. The Castle Martin, though small and black, did possess some white markings and had retained its milking capability as well as its ability to fatten at about three years of age. These cattle, along with the other Black cattle of Wales, were amalgamated to form the *Welsh Cattle Herd Book* in 1874.

A cousin of the Welsh Black, one local strain which had found a home in Ireland during the early Bronze Age, developed into the Kerry breed and from these a Mr Dexter, who was agent to Lord Hawarden on his estate in County Kerry, selected small animals and produced the well-known dwarf cattle which bear his name. Dexters were introduced into England in 1882 and have been kept by enthusiasts of the breed ever since.

Dairy breeds were taking on an ever-important role in agriculture, and with the corn slump in the late nineteenth century this selection accelerated. The first specialist milking breed in Britain was the Ayrshire, but the Alderneys, or Channel Island breeds, were also beginning to make their mark. Ayrshire cattle were developed by the expert stockmen in the south-west of Scotland for the production of milk to supply the dense populations of Glasgow and the south-west of that country.

The Dun cow of Suffolk, a breed which had been praised for its milk yield for the previous two hundred years, was transformed with the introduction of Norfolk and Dutch blood into the dual-purpose Red Poll. Due to this cross-breeding, it lost its milking capability and was allowed to die out. This act is said by agriculturalists to be one of the mysteries of agricultural history, and why this type of animal with so much potential was ignored is difficult to understand. The Red Poll is said to be a cross between the Red Norfolk and the Suffolk Dun, but this theory is not supported by all authorities as it seems likely that the blood of other breeds, such as the red Aberdeen Angus, may have also been used.

The Alderneys, or Channel Island, breeds were probably very closely related to the old Suffolk Dun, being of the same type and colour. These animals were being imported into southern England as early as the first half of the eighteenth century and were used at that time to improve the milking ability of the dairy type Short-horn. These is little doubt that some Alderney cattle were used as the foundation stock for the Ayrshire breed, although Whitehead again puts forward a suggestion in his book that some white cattle, probably the Chillingham breed, could be responsible for the white colour in Ayrshire cattle. Although looked on at first as the gentleman's cow, the Alderney was imported to England in vast numbers in the nineteenth century.

The capability of cattle as animals for traction purposes became less and less important and by the end of the nineteenth century, breeds had developed along three different lines; cattle for meat, for milk, and those of a dual-purpose type. The Longhorn had almost disappeared from the scene and the Shorthorn also began to lose its role as a dairy animal.

The trade in cattle from Holland was restricted, due to the outbreak of cattle plague in that country during the late 1880s. After this date, many of the animals which came into Britain from Holland were of the black and white, or Friesian breed which had originated in Germany and Denmark. It was these animals which became the foundation stock for the British Friesian breed, and made up the first *Herd Book* in 1911. Meanwhile, most other local types, such as the Gloucester, became unimportant commercially and reverted to breeds of fancy, retained only by enthusiasts.

Sheep

The dominant breeds of sheep at the beginning of the eighteenth century were, as we have seen, influenced by Bakewell's improved Leicester and most long and medium-woolled types owed many of their improved qualities to these Dishley

Wensleydale ewe with twin lambs.

animals. The Southdowns had been used extensively on the short-woolled breeds and had had a great influence on them.

The New Leicester was developed to replace Bakewell's type, being smaller and earlier maturing. It was directly responsible for the production of at least three new breeds. The Wensleydale, a cross between the old Teeswater ewe and the New Leicester ram, was capable of producing two or more lambs which, at that time, was relatively rare in other breeds and it became a fixed type in 1860. The New Leicester rams were used on the old unimproved Welsh sheep to produce the Lleyn type and the Border Leicester was a selected strain of the New Leicester which was developed in Northumberland by Matthew and George Culley. There also seems to be evidence of some Cheviot blood having been infused in the latter, but this is purely a matter of supposition.

As with cattle, at this time breed was less important than type of animal, and mixing of the breeds did regularly take place. The criterion was to select on type rather than on breed.

The oldest pedigree Border Leicester flock in Wales, owned by Mr L. Wrench of Hope near Wrexham.

By the middle of the eighteenth century Border Leicester rams were being used on Cheviot ewes to produce their famous Half-bred progeny.

Bakewell's Leicesters had first reached North America in 1799 and, as in Britain, they were popular for crossing with other breeds. In New Zealand they were responsible, along with the Merino, for the production of the Corriedale breed. The Dishley Leicester also spread and left their mark in every country in Europe.

By the middle of the nineteenth century, the Blackface had settled in the central highlands and islands of Scotland, where the grazing was mostly of poor grass and heather, and the Cheviot became popular on the better grazing lands of the south and north-east Scotland; but some small pockets of primitive sheep still remained along the west coast and the Western Isles, as they do today.

One breed of blackfaced sheep, which became recognised as pure in 1810, was the Suffolk. It had originated from crossing an improved Southdown ram with Norfolk Horn ewes and was itself hornless, black-faced and short-woolled. This breed is still one of the most popular for providing rams to use on almost any other breed of sheep for the production of early mating fat lambs.

The early maturing Southdown influenced most of the Down breeds of sheep by the end of the nineteenth century, but as a pure breed had, like the New Leicester, become unpopular except where it was used to cross with Upland ewes for the production of fat lambs.

The Shropshire Down was the result of putting a Southdown ram on several local breeds including the Morfe Common, and it emerged as a fixed type in about 1850, and was to become in the short term one of the most admired sheep of its time.

The Kerry Hill and Clun Forest breeds also possess some of that early Southdown blood and appeared at about the same time.

A Whitefaced Woodland ewe and lamb. The breed was much improved on the Chatsworth estate in Derbyshire.

By the middle of the nineteenth century, the Merino blood which King George III had brought to England sixty years previously had spread throughout many of the Downland sheep, but as a breed the Merino had fallen from importance. One breed, however, on which it left its mark was the Whitefaced Woodland breed of the Pennines. Merinos from the king's flock had been bought by the Duke of Devonshire in the late eighteenth century and dispersed to his tenants at Chatsworth as a means of improving the local type of sheep on the estate. Now classified as a rare breed by the Rare Breeds Survival Trust, the Whitefaced Woodland retained the hardiness of its ancestors, at the same time producing a better quality fleece.

Britain's hardiest sheep, the Herdwick, whose home is the Lake District where it is able to survive the high rainfall and poor pastures, was developed and improved still further at this time.

Another experimental cross which took place between the Scottish Blackface and the Hebridean resulted in the Boreray, a small breed which is now present on Boreray Island in the St Kilda group. It is of little agricultural importance at the present time, although some of its qualities may be of use in the future.

Pigs

Due to the reproductive rate of the pig, the import of animals in the eighteenth century produced a great variety of new types; Chinese and 'Neopolitan' pigs were crossed and re-crossed with the native European animals. The Berkshire pig was the first of the fashionable early breeds and was used to cross-breed with local types to produce many regional strains.

The Chinese pig was domesticated by the first neolithic farmers in about 3,000 B.C. and was descended from the wild *Sus vittatus*, a fatter, shorter-legged type than

Siesta! A litter of eleven Saddleback piglets (one month old).

the European *Sus scrofa*. The 'Neopolitan' pig was said to have been of Siamese ancestry and bred from the Asian pig, *Sus indicus*. The Siamese pig, or 'tonkey' pig as it was known, was also introduced into Britain direct in the eighteenth century together with a variety of other foreign types.

After nearly a century of cross-breeding of the various imported pigs, with the native *Sus scrofa*, regional types or breeds were becoming more discernible. The erect-eared White pigs of the north-east, which are said to have been the forerunners of the Yorkshire or Large White and which Bakewell used for his breeding trials some years before, were to become the most important. The black and white pig of Essex which was later to be called the Essex breed, the Tamworth which was the Staffordshire descendant of the old Berkshire type, the Hampshire black and white pig which later became the Wessex Saddleback, and the improved Berkshire all had a role to play in the pig industry of the time. The Essex and Wessex breeds combined in the middle of the present century, to become the British Saddleback. In 1820 the Wessex Saddleback was exported to Massachusetts and from it the American Hampshire breed of today was derived.

The Berkshire, first recorded during the eighteenth century, was at that time predominantly a sandy, red, spotted type, but during the early nineteenth century changed in appearance due to the work of Lord Barrington who, with the use of Neopolitan blood, created the black pig with white extremities and a dished face, very like the Bershire pig of today. It was renowned for its early maturing quality and the sweetness of its meat, which found a ready sale in the city meat markets.

The Tamworth, the only red native breed found in Britain, is thought by Robert Trow-Smith to be merely a strain of the old improved Berkshire which retained its sandy colour and built a reputation for itself in the Midland Counties. It also became very popular in both the USA and Australia early in the twentieth century and was exported in vast numbers.

Two more interesting, but probably untrue, theories regarding the Tamworth are that the breed was derived from a male, red jungle pig which was given to Sir Francis Lawley of Staffordshire in about 1800. It was allegedly used locally to mate with local sows and produced in the Tamworth area many red offspring, which became known as Tamworth pigs. Alternatively, it is said that Sir Robert Peel, of police fame, who had estates near Tamworth, imported a red pig, thought to have been a descendant of the Irish Grazer, and this animal was the father of the Tamworth breed. But as the Irish Grazer has never been recorded and no information about it has ever come to light, this seems to be rather an unlikely story. Robert Trow-Smith is probably much nearer the truth and his explanation seems to be much more feasible.

The Black Dorset pig was said to be a mixture of the various tonkey, Chinese and Neopolitan blood, and was produced in 1860 or thereabouts, by Mr John Coates of Blandford. The Gloucester Old Spot is said to be a strain developed from another mix of the different types of pigs and was produced in the middle of the nineteenth century in the Severn Valley. It was traditionally raised on whey from the cheese-making in that area, and on windfall apples from the cider apple orchards.

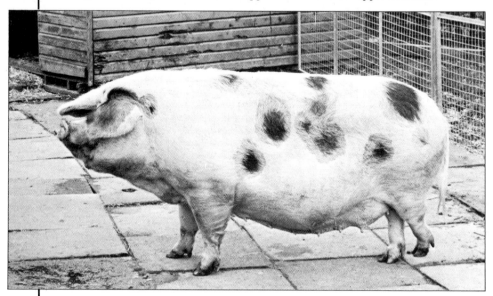

An in-pig Gloucester Old Spot sow.

Pig breeds have only in the past half-century become of interest to livestock breeders. They are all of recent origin and are of less importance commercially than the type of pigs to be found within the breed. This theory is borne out by the present-day development of strains or hybrids rather than breeds.

Horses

Horses of the nineteenth century had been more or less developed during the previous hundred years and breeders found little more work to do, except perhaps a few modifications to secure a more uniform type. The heavy horse breed which was common on farms at the end of the nineteenth century was the Clydesdale, which had originally been bred by using a Flemish stallion on large pony mares, but also is thought to contain a certain amount of Shire blood which had been infused in the late seventeenth century. The Shire itself, which had been bred from the old English Black horse and had the benefit of being the subject of some of Bakewell's experiments, was developed and improved in type by the end of the eighteenth century.

The Suffolks of the late eighteenth century were said to be descendants of one great sire who stamped the breed with his characteristics. The only other agricultural work horse of the time was the slightly less heavy Percheron, a grey horse developed

A young stud Shire stallion at Croxteth Park, Liverpool.

in Normandy. This breed, too, was, like the British horse, a descendant of the Flemish type, and it became more popular in America than in Britain.

By the end of the century, British produce was facing competition from abroad. With the new method of refrigeration, meat was brought from countries at the other side of the world, from Australia, New Zealand and from South America. Home-produced butter, cheese and meat were facing rising competition both in price and quality from the imported article.

The stage was set for a change in the scene of the endless drama of British livestock farming. The dairy cow was to gain the important place which it held for the next three-quarters of a century and this was to be modified only by the effect of two World Wars, and the need to produce as much of our own food as was possible in these islands.

7   The Twentieth Century

Cattle

At the beginning of the twentieth century, most cattle in Britain were still referred to by regional names, as well as their physical characteristics, for instance the Cheshire Longhorn, or the North Country Shorthorn. There was a great variation from district to district, and as one regional type became more popular so it spread to other districts. The Hereford, North Devon, South Devon and Sussex, together

A Dairy Shorthorn. The breed is not naturally polled, but in commercial herds calves are de-horned at about one week old.

with the Shorthorn and, to a lesser extent, Longhorn and Red Poll, were all breeds of England. The Aberdeen Angus, Galloway, Highland and Ayrshire breeds were all of Scottish origin, and the Welsh Black (as explained in Chapter 6) was an amalgam of all the Black cattle from the Principality, which by the turn of the century were classed as one breed. The Alderney cow from the Channel Islands had sub-divided into the Jersey and Guernsey, and the Kerry from Ireland existed in its normal form and also in its dwarf offshoot, the Dexter.

The dairy cow had become a specialist milk producer, and breeds which made up this group in the first decade of the century were the Dairy Shorthorn, which outnumbered all other cattle, the Ayrshire, Jersey, Guernsey, and to a lesser extent the Longhorn, South Devon, Red Poll, Kerry, Dexter, and a new addition to the list

An in-calf South Devon cow. Note the particular whiteness of the horns.

A first prize Aberdeenshire (early Aberdeen Angus) bull.

which was beginning to gain some popularity, the British Friesian or Holstein.

In the Shetland Isles, the thrifty, indigenous breed retained a hold and other minor regional types, such as the Gloucester, were kept in small groups. Claims were made that some of the breeds served a dual role, providing both milk and meat. The Red Poll and the Welsh Black were typical examples, but the supreme animal in this group was the Shorthorn.

Other breeds made no claim to being dairy animals, their sole purpose being that of producing quality beef. The North Devon, Sussex and Hereford were meat animals, the latter breed being by far the most numerous.

Scotland and the North Country boasted the Beef Shorthorn, Galloway, Aberdeen Angus and the Highland.

A White Galloway with dark points at Cholmondeley Castle, Cheshire.

Before 1920, the Shorthorn was the most widely distributed breed, both at home and abroad, and it was said by breed enthusiasts to have risen in popularity due to its unrivalled 'range of adaptability, thriving in all countries and climates'. Three-quarters of all dairy cattle in Great Britain and Ireland were Shorthorn, and the breed was said to have the ability of stamping its good character on all its progeny. It was also quoted as being 'more important than any other breed whether it be viewed as a grazier's beast or a dairyman's cow, and it may be seen at nearly all fairs and cattle markets in the country, a statement that can be made of no other breed.'

The Longhorn had fallen from favour to a marked degree at the beginning of the twentieth century, but the few breeders who stuck by the breed decided by selection to concentrate on a red, or brindle-coloured, animal, with a white stripe along the spine. The majority of cattle that had survived were now of this one type, with a common colour and markings. The old Longhorn of which each area had its own type, was replaced by the new Shorthorn. Those that did survive were true dual-purpose cows and in some remote areas still produced, where they were required, working oxen.

The Hereford was said to be immune to tuberculosis, able to stand drought better than any other breed, and command the top price in the London market.

Of the two Devon breeds, the South was the dairy beast and gave a good account of itself as a beef animal. It was claimed to be the heaviest breed of cattle known. During the late 1890s the breed was exported in large numbers to South Africa, South America and Australia. The North Devon was purely a beef breed and no claims to its milk potential were made. Both these breeds were red in colour. The Sussex was very like the North Devon, being a beef producer.

Photograph of a Shetland cow before the introduction of Friesian blood to the Island, circa late nineteenth century.

The Red Poll, which had been produced during the nineteenth century by crossing the Norfolk Red and the old Suffolk Dun, was the only naturally hornless English breed and was a dual-purpose animal.

The Welsh cattle retained their hardiness and dual role, and were animals of the poorer land, but they still provided store cattle to fatten on the better English pastures. They, too, were relatively free from tuberculosis.

The Scottish polled breeds, the Aberdeen Angus and the Galloway, which had been selected for their beef qualities during the nineteenth century, underwent still more selection at the turn of the century and their milking qualities were lost almost completely, leaving them with the ability to provide only sufficient milk to rear their own calf. The beef which they produced was of the highest quality and was in demand at all the meat markets. The Galloway, unlike the Aberdeen Angus, which was always by this time black, had a range of coat colours, through dun, red, brindled and black. The latter colour was more favoured, especially in England. The Belted strain was only popular as a fancy, but this characteristic which was carried by the recessive gene has, until today, retained a small number of enthusiasts.

The Highland changed little and this was solely due to selection within the breed.

The type had probably remained the same for the best part of ten or twenty centuries. It thrived where no other cattle could exist and produced beef of high quality. The milking potential had never been high, but may have been exploited only by crofters for a few weeks after calving.

On the Shetland Isles the indigenous cattle were kept mainly by crofters, for producing both milk and a grazier's calf from the poor pasture of the Islands.

The dairy breeds were rising to the challenge of still higher production to supply not only dairy products, but the increasing demand for fresh milk. The Ayrshire, the

A modern Shetland cow.

only true British dairy breed, was the leader of the fields for milk production, partially replacing the Shorthorn as a specialist producer, first in its home region of south-west Scotland, and later in both Wales and England, especially around London, and also in Northern Ireland. The breed was exported to all parts of the world, particularly the United States, and it was said that 'no cow will yield a greater return for labour bestowed upon her'.

Channel Island breeds, which had been kept pure in their respective islands, due to the strict import regulations instigated in 1763 which stipulated that no other cattle be imported to the Islands, were popular in Britain due to their yield of rich milk.

The Kerry, cousin of the Welsh Black, had been selected along different lines in Southern Ireland and was solely a dairy cow which was capable of living and producing from the poorest of land. Its offshoot, the Dexter, provided breed enthusiasts with an economical yield of both milk and beef, and for its size was said to be a hardworking and efficient animal. It was reported to throw a very useful calf when crossed with an Aberdeen Angus bull.

The Irish Moiled, which is a polled medium-sized dairy breed with a white back stripe, was to be found in the north of Northern Ireland in the early years of the

An Irish Moiled bull calf at Croxteth Park, Liverpool.

century. It was said to be an economical breed to keep on the poor land of that area, for the production of a high milk yield. It was also said to produce a useful store. The breed fell out of favour, however, and is now represented by under 100 individuals. The Rare Breeds Survival Trust has taken an interest in the breed and hopes to save it from becoming extinct.

The Friesian, or Holstein, had arrived on the agricultural scene at the end of the nineteenth century, but had made little impact in the first decade of the twentieth.

Sheep

The sheep breeds of the early twentieth century were, like cattle, distinguished by their names being taken from their region of origin, or as in the case of the Blackface a physical feature.

In Scotland, the main breeds were the Border Leicester, Cheviot, and the Scottish Blackface. Wales had its Mountain sheep, and the Kerry Hill from the Border Country. The Irish breed, the Roscommon, although not widespread, still had a role to play. In the Western Isles and the Isle of Man the primitive breeds survived in their own longstanding domains, but only in small groups. Manx Loghtan, the Hebridean, the Shetland, the seaweed-eating sheep of North Ronaldsay, and the longest survivor of all, the little Soay, retained their rather tentative hold.

The breeds of England included Mountain breeds, such as the Herdwick of the Lake District, the Lonk and Swaledale of north-east Lancashire and the neighbouring counties, and a few survivors of the Whitefaced Woodland breed. There were also some regional types such as the Exmoor and the Derbyshire Gritstone. Most of the regional Longwool breeds still existed: the Leicester, Lincoln, Cotswold, Devon and Wensleydale. The Shortwool types included the Suffolk Down, South-down, Oxford Down, Shropshire Down, and Ryeland breed, among others.

A well-bred trio of Manx Loghtan sheep. Note the well-shaped four horns on the ram; the ewe and lamb are both two-horned.

The Suffolk was the popular breed for crossing with Longwools to produce a good carcase of lean meat, a good quality fleece and good mothering qualities in the offspring. Border Leicester rams were used in flocks of Cheviot ewes to produce the famous Half-bred, which was then put to one of the Downland breeds to produce a prime, fat lamb.

A Swaledale ewe, showing the typical curled horns of North Country blackfaced mountain sheep.

Bakewell's Leicesters, having been further selected, also resulted in one of the most popular breeds for crossing. In the early twentieth century, one Bradford wool specialist was of the opinion that the Leicester/Merino cross produced the 'crossbred clip of the world'. At the Smithfield Club in 1907, Leicester sheep were awarded the championship for Longwools of any breed.

The Border Leicester, which resulted from the work of Bakewell's students George and Matthew Culley, was given its distinctive name in 1869. The Kent (or Romney Marsh), due to its adaptability in marshland conditions and its sound constitution, was exported to North and South America and Australia, where it experienced times of year when food was abundant and times when food was scarce, much as it would have done in its original region in Britain.

The Down breeds retained their position in their respective district of origin, the Oxford being the heaviest and thriving well on both arable and grassland. The Shropshire retained its popularity for, amongst other reasons, its high prolificacy, with a lambing percentage of between 150 and 175 per cent.

The Ryeland, during the first part of the century, became a larger animal, giving a much heavier wool clip, but it still retained its sound constitution and was still popular for its ability to thrive on cold, damp soil. Ewart, speaking of the Ryeland in 1837, said 'it would endure privation of food better than any other breed'. Sir Joseph Banks had said 'the Ryeland deserves a niche in the temple of famine.'

During the first half of the twentieth century, wool production was of great importance, but the production of meat was more so. The breeds of sheep were very numerous, each type adapted to its own area of origin and the specific conditions which were to be found there. It was the adaptability of the animals to survive under different conditions to be found in the various types of sheep which was the deciding factor in the choice of the breed. As has been shown, the Longwool, or Down breeds, could not survive on the hill land, therefore a more adaptable breed which was better able to tolerate the environment would be selected. Adaptability is the sole reason why so many different types or breeds of sheep had survived for so long. Due to the varying conditions to be found in the British Isles, from the mountains of Scotland, the Lake District and Wales through the flat, heavy soils of the Midlands and eastern counties to the chalk downs of the south of England, these indigenous types of animals had evolved.

Pigs

Pigs, at the beginning of the twentieth century, had more or less developed from regional types into breeds. The Large White had advanced in public favour more than any other breed. It had prick ears and its lack of colour produced a more acceptable product, particularly bacon which was said to sell more easily. Large White boars were taking on the role of sire to all other breeds, as their offspring, even if produced by coloured sows, were mostly white in colour. A typical example is the Large White boar crossed with a Saddleback, or Large Black sow; the off-spring are mostly white with just a few patches of blue. The Large White was also well liked for its adaptability. It could be slaughtered young to produce the best quality pork, left to grow a little longer and produce first-class bacon, or allowed to grow fat and heavy to give a carcase of up to 500 lb (227 kg) deadweight, which

A Lincolnshire Curly Coated pig. The last pure pig of this breed died in 1972. Croxteth Park are endeavouring to re-build the breed.

during the early years of the century was popular in industrial areas. The sows of the breed were known to produce a good litter, ten pigs on average, and these would thrive and grow well due to the large quantity of milk produced by the mother. The only disadvantage of the breed was that of becoming sunburned, due to the skin having no colour pigmentation, but this was of minor importance and did not prevent the breed from becoming established over the whole of the country.

The Middle White was a shorter, more early maturing animal, with a short, dished face. It was popular as a specialist producer of pork, or if left to grow provided a smaller, heavier fat carcase than the Large White. Boars of the Middle White breed were exported to Russia, America, Australia and Japan, where they proved to be very popular for crossing with other breeds.

The Small White was also popular in Britain at the turn of the century and produced a higher proportion of fat to lean in its meat.

The Berkshire and Tamworth breeds had developed into the types which we know today. The former was black in colour with white feet, white tip to its tail and white on its dished face, but at this time had a restricted distribution. The Tamworth had become free of black blotches in its coat and was gold/red in colour, said to be due to the introduction of some Yorkshire strains of Large White blood. It also became leaner and more active. It was widespread in the Midland Counties where its hardiness and ability to thrive in all weather conditions made it very popular. The breed was also exported to Australia and America in large numbers during the first decades of the twentieth century.

Josephine, Sheffield Metropolitan district Council's much admired Middle White sow.

The other coloured pigs, the Large Black and Saddleback, had a role to play in outdoor pig breeding units, together with the Gloucester Old Spot. All three breeds were hardy, prolific and, if crossed with a White boar, produced almost white pigs which were popular with fatteners at that time. They were docile and not apt to stray, due to their large flop ears which prevented them from seeing too far ahead. They were also able to tolerate the sun more easily than the pure White pigs.

Three-week old Middle White piglets. Note the noses beginning to turn up characteristically.

The Gloucester Old Spot of that time was covered with many more spots than it has today, therefore giving it the advantage of the coloured pigs. One exception to this rule was the Lincolnshire Curly Coated pig which was pure white, but as its name suggests had a long, curly coat which also gave it a certain amount of protection from the elements. It, too, had flop ears and was said to be as prolific and hardy as coloured pigs and as early maturing and efficient at production as White pigs. The Lincolnshire Curly Coated was well-known in the east of Lincolnshire at the turn of the century, but spread much further afield and became more popular in the early 1900s.

Horses

Work horses of the early twentieth century were still the Shire, the Suffolk Punch and the Clydesdale. The largest and strongest was the Shire. It weighed up to 2,000 lb (900 kg) and was as high as seventeen hands or more. By this time, it was mainly bay or brown in colour, but black and also white were still to be seen occasionally. It had long hair, or feather, in the lower part of its legs and a long mane. Although the breed was so strong, it was also renowned for its intelligence and docile temperament.

The Clydesdale was a smaller animal, being up to sixteen and half hands in height. Its colour was brown, black, or less commonly grey, and often the coat had a dappled appearance. The feathering on the legs was much less than that of the Shire. Many animals of this breed were exported to America and the colonies in the first decades of the century, and in 1911 for instance 1,617 animals were sent abroad.

The Suffolk was shorter than the Shire and the Clydesdale, but up to 2,400 lb (1090 kg) in weight. It had no long hair on its legs and was known as a clean-legged horse. The colour was always chestnut, but slightly different shades did occur from time to time; this regular marking is said to be a striking testimony to the purity of its breeding. The breed was very popular in both America and the Colonies, where it was used to cross with native mares and, as with the Clydesdale, many animals were exported in the early part of the century.

The 1930s and World War II

In the 1930s changes in agricultural practices began to accelerate; not only were they dramatic but they also took place faster than had ever happened before. During the first decade of the century progress had been constant but steady. With the introduction of organisations such as the Milk Marketing Board and the Pig Industry Development Authority, patterns and traditions took a change of direction. At that time British farming was mainly mixed. Roughly 46 million acres of land were available for animals and other food-producing crops. There were approximately half a million separate farmers who employed about one million other people. At the present time, the number of farmers has dropped by half and they probably employ about half a million other people; and 50 per cent of the total land is now made up of large specialist units.

The British Isles produced at that time about a third of the total food needed for home consumption and every farmer had a livestock enterprise of one form or another, and in the north and west livestock were predominant. These animals, each with its own special characteristics, were developed in every region. Weather conditions, soil type and local needs all helped in the selection of these well-adapted creatures.

It was the beginning of World War II which saw other pressures arise and these were to dominate the direction which agricultural production would take for the following half century. Farmers were given economic signals and these had the effect of over-ruling all other criteria in the development of farm livestock. For the working horse this heralded the final chapter in its role on British farms. With the introduction of the small agricultural tractor in place of the giant steam engine, the work of the agricultural horse was taken over by the internal combustion engine, quickly to be followed by the diesel or compression engine. Harry Ferguson, the inventive genius from Northern Ireland who gave his name to the most well-known make of tractor, was instrumental in this agricultural mechanical revolution.

During the war, arable acreage increased by over 50 per cent and more food was grown in this country than had ever been produced before. The economic signals directed agriculture to the new policy of self-sufficiency in food production and technology transformed the whole nature of farming. This policy, which had been a necessity during the war when our food supply line was almost cut by the German navy, became permanent until Britain joined the European Economic Community in 1973. From that date onwards, agricultural decisions have been taken in Brussels.

At the end of the 1930s Britain had entered the war, and though breeds of livestock had changed due to more scientific breeding and a different form of selection, many of the traditional regional breeds remained, albeit in some cases in very small numbers. The Friesian and the Ayrshire shared the role of national dairy cow, with the Dairy Shorthorn, but other breeds were still to be found all over the country. Herds of different-coloured cattle grazed in fields from Land's End to John O'Groats. Sheep found on the hills had changed little, if at all, but some of the lowland breeds such as the Cotswold and the Shropshire Down had declined in numbers. The arable system of sheep farming, which had been practised on the chalk and limestone uplands of the South Midlands, East Anglia and the Yorkshire Wolds, declined and the breeds which were an integral part of this farming system, and which were unique to it, also dropped rapidly in numbers. A few flocks remained but they were essentially of pedigree status and specialised in the production of rams which were bred for the sole purpose of crossing with upland sheep to produce a better, early maturing lamb. The only area of lowland which remained, as it does to this day, heavily stocked with sheep was the Romney Marsh. As farm animals sheep have, over the past fifty years, been relegated to land which cannot be put to any other use, or have become part of a secondary enterprise. Even today, however, no other country is able to rival the variety of Britain's sheep types and up to sixty pure breeds, together with many more regional types and cross-breeds, are to be found. The minor breeds which have retained their place in the regions of their birth still have a role to play and have characteristics which are special to them and may one day find a more important place, not only in Britain's economy, but also in that of other countries.

Postwar Developments

Different breeds of pigs could be seen on farms up and down the country at the beginning of the war in 1939. The Gloucester Old Spot, the Wessex Saddleback, and the Essex as well as the Middle White and the Lincoln Curly Coat, were all familiar names in agricultural circles. As new technology developed, new systems of husbandry were born. The structure of the pig industry took a completely new direction which was due to its great sensitivity to the cost of feeding stuff and to market requirements, and it steered consistently towards an industrialised system of production. The new method of pig farming saw the beginnings of the demise of traditional pig breeds. Slaughterhouses and pig processing plants were beginning to require animals which could be diverted from one market to another in order to achieve the most profitable outlet. By 1950 the changes that had been instigated by the needs of a Britain under siege during the war years continued. Self-sufficiency was the aim of each of the post-war governments and to this end agriculture was given financial incentives to obey government directives.

Milk production was encouraged and, to a lesser extent, so too was the production of beef, sheep and pig meat. Arable farming which had, until the war years, been in a depressed state took on a new lease of life. Land which had been used for outdoor pig enterprises or sheep-folding was now set aside for growing arable crops for human consumption. Pig farming, as well as poultry farming, became intensive and farms became like factories. Dairy units, too, became more and more specialised and with the use of artificial fertilisers, especially chemical nitrogen, the stocking rate of land increased. In 1930 between three and five acres, depending on the type of land, were needed to keep one cow, but by 1980 this could be achieved with one acre of reasonable land.

With the introduction of artificial insemination (see Appendix I) by the Milk Marketing Board for cattle, and the Pig Industry Development Authority (later the Meat and Livestock Commission) for pigs, the popular and most economical breeds of the day could be used as herd sires by most farmers. Fifty years ago there were over twenty breeds of dairy cattle, but now over 90 per cent of the total milk production comes from one breed, the black and white Friesian, or Holstein; and during that period even this breed has changed from its original type into a bigger animal capable of giving very high yields of milk. This transformation has been almost entirely due to the use of artificial insemination. The cow of today has lost many of the qualities of the old breeds. It is cossetted and managed to a very high standard and in some cases is kept under a system called 'zero grazing'; it remains inside a yard all its milking life and is never turned out into a field to graze. Grass is cut and mixed with other produce before being fed as a complete diet. In some herds, due to the large yield, the cows need to be milked three times every twenty-four hours, instead of the usual twice daily.

The trend in agriculture from the war years onwards has been to shed labour and to intensify production. To cater for this system, different types of livestock, especially pigs, cattle and poultry, which would tolerate the new practices, have been developed. Sheep were little affected generally, especially on the hilly land where little more could be achieved, except perhaps some selection to improve the existing types. To a small degree, some foreign breeds of sheep have been imported recently for the purpose of developing characteristics to cater for present-day requirements. The Texel, from the north of Holland, and the Finnish Landrace are two such examples. The former has been used mainly to cross-breed as a flock sire to improve the growth rate and lean meat quality of its offspring, whereas with the Landrace the objective is to pass on its prolificacy and milk-producing capability to some British breeds with which it has been crossed.

From its inception in 1933, the Milk Marketing Board of England and Wales, together with other similar organisations in Scotland and Northern Ireland, has stabilised the dairy industry, not only in its first objective of marketing all the milk produced, but also in providing a means of breeding and selection of both dairy and beef cattle. The dairy cow by 1950 had risen to be the corner-stone of British agriculture and this pinnacle it held until 2nd April 1984. On that date, the introduction of milk quotas enforced by the European Economic Community brought about a halt to the ever-increasing production of milk, the result of which we have yet to learn.

Arificial insemination minimised the transfer of diseases between cattle. It also gave even the smallest dairy farmers the opportunity to use semen from some of the best bulls in the country, thus bringing about swift improvement and standardisation to the national dairy herd, but at the same time the great majority of cattle became stereotyped. The wide variety of strains which had previously existed were overtaken by the new fashion which catered for the short-term economic needs of the time.

With the introduction of recent legislation, a totally different economic pattern may now emerge and some of the special characteristics of breeds other than the Friesian may once more be required. A dairy cow capable of producing milk from poorer quality feed, kept on a lower input, lower output system, may once more become an economic unit. Artificial insemination could once again bring about a very swift change in Britain's national dairy herd if semen from other genetic types is available. Cattle semen can be frozen and stored at minus 196 degrees Centigrade in liquid nitrogen for many years. It is, therefore, possible to delay the use of a bull until his value according to the needs of the time have been assessed. The semen, too, can be diluted, and one recent record set up by a Milk Marketing Board bull provided 80,000 inseminations in one year (1983).

The Milk Marketing Board also developed the practice of producing beef from the dairy herd. Once more, artificial insemination enabled dairy farmers to use different breeds of bulls on selected cows. For the top producers in a herd the semen from a dairy bull may be chosen and used to provide replacement animals, whereas on the poorer animals insemination with the semen from a beef bull could be used to produce a calf which was solely intended for meat production. For the latter exercise originally only the British beef breeds, such as Hereford, Beef Shorthorn, Aberdeen Angus and Galloway, were used, but as public taste changed and more economical, leaner cuts of meat were demanded, the Milk Marketing Board imported bulls of the continental breeds with a view to obtaining a better beef calf from the pure dairy cow.

The Charolais, an animal from Central France, whose herd book had been established at the end of the nineteenth century, was imported into this country. The breed had retained its ox-like physique and was capable of rapid growth and the production of lean meat and muscle, without laying down fat. Twenty-six bulls were imported in 1961, followed by many more importations. The British Charolais Cattle Society was founded in 1962. Other similar beef-type breeds such as the Simmental and the Chianina, were also brought from Europe. These continental breeds have proved very successful for crossing with the finer-boned dairy cattle such as the Ayrshire and Channel Island breeds to produce beef calves, but overall the Hereford has remained the most popular beef sire, especially for crossing with the black and white breeds of cattle. A polled (hornless) strain of Hereford has now been developed, capable of producing a high quality beef calf which is also polled.

The Friesian cattle imported into this country from Holland during the middle of the century were of a dual-purpose type, being capable of leaving a good carcase when they came to the end of their useful milking life. They also had the added advantage of producing a top-class rapidly growing steer. During the last decade, however, another more dairy-type Friesian has been imported from Canada and is referred to as the Canadian Holstein. This strain carries much less flesh than the animals which were brought from Holland and a new herd book has now been established. The two strains have, in many cases, been cross-bred and very few commercial herds contain either type in its pure form.

The beef breeds of cattle at the present time can be divided into two categories. One contains pedigree animals which are retained almost exclusively to provide herd sires used in the production of beef calves from both dairy and suckler herds. This group includes the Hereford, the Beef Shorthorn and some Continental breeds. The second category contains the Scottish breeds, such as the Highland, Galloway and Aberdeen Angus, but also to a lesser degree the Hereford. Cattle in this group are still to be found in a pure, if not always pedigree, form in the region of their birth, still serving the purpose for which they were originally intended.

A relatively new technique which has now revolutionised livestock breeding, especially that of dairy cattle, is the transfer of ova or embryos. The process can be used on most domestic animals and involves the transfer from one female individual to another of fertilised eggs, or embryos. The transplantation is carried out quite routinely and successfully with dairy cattle and, although surgical methods are still used, non-surgical means, very much like those used in artificial insemination, are now being developed. The major benefit gained from embryo transplanting is the capability of a superior animal to produce a number of offspring in any one year, instead of the one usually born by natural methods. Once again, however, the process together with that of AI results in many animals being produced which are all related and are of the same genetic type. Though these animals are perhaps ideal for today's conditions, they may not be at all suitable for the next decade when economic pressures may be exactly the opposite to what they are at the present time.

Pig breeds have now all but disappeared as Britain is required to produce more and more pork and bacon from intensive pig farms. Two breeds now account for 90 per cent of all pigs: the Large White whose ancestors Bakewell used in the eighteenth century, and the Landrace which was first imported from Sweden in the 1950s. At the present time 70 per cent of all sows used for commercial production are crosses between these two breeds. Both are used pure, or crossed one to another and by so doing have the advantage of hybrid vigour. There is a great deal of variation within the same breed and some strains show very different characteristics. The Welsh breed which is now very much like the Landrace, due probably to an infusion of blood from the early importations, is also used to a certain extent, but the British Saddleback made up of the old Wessex and Essex breeds, the Tamworth, and the Gloucester Old Spot, together with the Middle White and Berkshire, are kept on a very small scale, whereas the Lincolnshire Curly Coated breed has died out altogether.

Artificial insemination has also had a part to play in the make-up of the modern pig, much as it has had in the cattle breeding programme. The practice was first set up as a side-line by cattle AI centres, but by 1962 the Pig Industry Development Authority (PIDA) later to be absorbed by the Meat and Livestock Commission (MLC) provided funds to these centres to enable more work to be carried out. In 1964 Dr Hugh Reid set up the first PIDA AI service which enabled pig semen to be delivered to individual pig farmers in many parts of the country. The one problem which has been found is that pig semen, unlike that of cattle, is more difficult to store. At the present time, whereas 80 per cent success is achieved with fresh semen, only 40 per cent is as yet possible with frozen. Another modern practice which is today used in pig husbandry is that of setting up minimal disease herds. Piglets are taken by Caesarean section from their mothers and reared artificially. Thus they never have contact with their own kind except other animals so produced. In this way they do not contact or carry any harmful organisms and can be used as the basis for the establishment of a disease-free pig unit. Being free of disease, they do not develop any natural resistance and are always at risk from infection.

The modern pig has been developed along the same lines as that of the modern dairy cow and is a specialist producer solely for today's needs and to suit today's conditions. It has the same narrow genetic make-up and would be of little use if required at some time in the future to retake its position as an animal which was capable of surviving on by-products and cheap food.

The farm horse, like the goat, has virtually disappeared from the farming scene and is now relegated to be classified as 'other livestock'. In 1939 three-quarters of a million working horses were used on British farms and almost every one of the half million farms had at least one horse. By 1970 there were only several hundred, and today heavy horses have become a rare sight; the few that now remain are primarily for show or display purposes and have become a hobby or fancy. Some are still used by firms such as brewers for short hauls where frequent stops are necessary.

Although agriculture has developed new techniques, crops and animals, people in some parts of the world still die of starvation. What the future holds for mankind is, to a great extent, dependent upon ourselves. It is our responsibility to see that we do nothing to destroy that which may one day be of benefit to future generations.

8 The Future

Man has the power to destroy with ease in a few moments what has taken in some cases millions of years to create. We take pride in the fact that with modern technology we have the means at our disposal to change our planet and so improve our existence. But we also have a destructive capability which, unless great care is taken, will not only destroy us but also our world. Sir Peter Scott has at the Wildfowl Trust Reserve, Slimbridge, a strategically placed mirror, where all visitors are sure to see it. Above the mirror is a caption, 'Look into the frame below and you will see a specimen of the most dangerous and destructive animal the world has ever known.'

From the end of the Ice Age our ancestors selected animals and plants which could be controlled and used to supply all of their needs. From that time, as we have seen, selection has taken place and this will continue to be so as long as man has free will and uses it wisely. From the end of the nineteenth century, changes have been swift and in some cases very dramatic. New discoveries have brought about benefits to mankind in a changing world, but those benefits have not been available to the whole of the human race. Whereas we in the West have benefited not only from the kindness of our climate but also from the efforts of our predecessors, the Third World and less advanced nations have received no such benefits. Such things as a plentiful supply of food, which we take for granted, is beyond the wildest dreams of a large section of the people of this planet. While we now produce food surplus to our requirements, they starve to death for the lack of it. Do we have the right to destroy what, to us at this moment in time, seems to have little role to play in our world? Though we have this power, we do not have the far greater power to recreate that which we have, through ignorance or neglect, caused to be lost for ever. Examples are to be seen every day of old practices or methods which many years ago fell from favour, only to be re-tried and found to be an improvement on, or a beneficial addition to, modern ideas.

With our new chemical sprays we could easily have destroyed many of the plants which are now used in modern medicine. If their benefits had not been rediscovered before the end of the next decade, new technology may have deemed them useless and allowed them to disappear for good.

This picture is similar with the old farm livestock. The regional breeds were developed over many centuries to serve a specific purpose. Many of them cannot compete with new strains or breeds at the present time. On the other hand, others are, like the plants, being rediscovered and taking on an increasingly important role in modern agriculture. Fashions and conditions change and during the last decade, for instance, pig farming took a new direction. The pig became the specialised producer of products which were in fashion. To cater for that fashion, the pig became a creature of a controlled environment, protected from the outside world and fed on an expensive, specially formulated diet. This change happened almost overnight and the pig which had lived outside and foraged for a living has at this moment very little part to play in the bacon and pork producing practices of today. The food which these modern, pampered pigs require could be used direct to feed human beings. The competition for food from an ever-increasing world population could easily lead to this situation. If this theory was, in fact, to turn out to be correct, the pig of today would be of no use. It could not survive the rigours of living like its ancestors and would die out. The old-fashioned breeds which can make use of by-products such as whey, fallen apples, waste of all sorts, grass and pannage, could once more become an important converter of products which are of little or no use if fed directly to human beings. The Gloucester Old Spot, Tamworth, Saddleback, Berkshire and Large White, would be set to take the place of the hybrid X,Y,Z.

The speed with which cattle, especially dairy cows, have changed in the last decade or two, has been critical. New methods of husbandry and complicated economic pressures, together with the effect of two World Wars, were major influences in the development of the cattle of today. The modern dairy cow has been turned into an animal capable of producing large quantities of milk and to a lesser degree meat by its being managed in a very specialised way, and fed on a very specialised diet. This has led to a surplus of milk being produced. Cows of today, like modern pigs, need to be pampered and cossetted and their dietary demands are also alike in many ways. The grass which makes up a major part of the diet of cattle is grown by using large amounts of nitrogenous fertilisers, on land which could more economically be used to grow crops for direct feeding to human beings, as opposed to being fed indirectly through the cow, in the form of milk and meat. This change in dairy cow type has led to a massive decline in the old breeds and a huge increase, as we have seen previously, in the popularity of the black and white Friesian and Holstein breeds. Cattle capable of living and producing milk and beef from other than high grade land have become unfashionable, the practice of today being high input, high output. Breeds such as the Shetland, which is capable of producing an economical return in the form of both milk and beef from the poorest land, nearly became extinct and was only saved in the nick of time. The Kerry and the Gloucester were also rescued at the eleventh hour. White Park and British White cattle, which have been associated with man for so long and were instrumental in the make-up of many of our domestic breeds are now thought to be an economic producer on land of poor quality but they, too, went along the same path as the previous breeds.

The specialisation in modern sheep farming has done much the same to breeds of sheep, although the mountain breeds have retained their hold on their traditional areas. The Ryeland, for example, was high on the danger list until recently, but is now experiencing a commercial revival.

Many people ask what is the point of preserving breeds of livestock which, at this time, have no obvious use. They say selection has produced animals capable of playing their part in our modern agriculture and as conditions change so the animals will change with them. This argument would have been justified if it had been aired even half a century ago, but now it can hold no credence. The unique characteristics which were acquired over the centuries by breeds of farm livestock are in grave danger of being lost forever due to the concentration of interest in breeds which only favour today's conditions at the expense of all others. Until the early part of this century, there were always to be found pockets of livestock which could be used to provide a reserve of genetic characteristics. Due to economic and political pressures this is no longer the case. The diversity of type is disappearing fast, to be replaced by a stereotype. Some characteristics contained by the old breeds may one day become vital for future farming systems, not only in Britain but in the developing world too.

Over the past two years in the Western world changes in the human diet have been recommended by medical and dietary experts. Lean, firm meat and lower fat dairy products are now preferred. Milk quotas introduced in 1984 have also resulted in a change in direction for the dairy industry. Some of the old breeds of farm livestock may now fill the needs of today's farmers and consumers. While there is still time, even though it is late in the day to start, there is still a chance to save some of these old breeds for future generations.

One charitable organisation in Britain has taken on the gargantuan task of rescuing these animals from the threat of extinction. The Rare Breeds Survival Trust Limited was established in 1973 and under its Hon. Director Michael Rosenberg is devoted to the conservation, study and promotion of Britain's lesser known breeds of domestic livestock. It publishes a monthly journal, called *The Ark*. Members of the Trust comprise farmers, livestock breeders, naturalists and enthusiasts, as well as museum staff; in fact anyone who has an interest in animals and Britain's heritage is welcome to join. Large national companies are corporate members, and through their help, together with the help of individual members, the Trust is able to carry on its work of protecting our livestock heritage and farming future.

The Rare Breeds Survival Trust Limited gives the following description of rare breeds. 'Until comparatively recently most breeds now considered rare were both numerous and widespread. Rapid changes in farming practice, fickle consumer preference and short term economic consideration are the main reasons for these breeds having become rare. Many of the breeds now in danger played an important part in the development of our livestock industry and have a role to play in its future. In the rarest breeds there may be only a few hundred or less individuals, perhaps in only a few herds. The qualities of most breeds have never been fully researched or recorded, so that only now are we beginning to assess their value in different conditions and how they could be of value in the future. The need to preserve these rare breeds is universally acknowledged by livestock breeders world-wide. Some breeds, until recently high on the danger list, are now experiencing a commercial revival. Others are shown to have a great future on marginal land. Who can tell what our needs and those of the rest of the world might be in the years ahead? Whatever they are, they can only be met from the gene pool of existing breeds. It is this future that the Trust seeks to safeguard.'

In the United States and Canada, the Trust equivalent is the American Minor Breeds Conservancy, which was founded in 1978, and this organisation has the same high ideals. Ridgeway Shinn, one of the founder members, says: 'Until fairly recently, the rural landscape of America was rich in a diversity of livestock breeds. Many regions developed specific breeds for the climate, topography, needs and tastes of the people of that area, but as American agriculture became more indus-trialised it developed highly specialised animals. Now a handful of breeds dominate the scene, while the rest are threatened with extinction. For instance, a decade ago 10,000 Tamworth pigs were registered annually in the United States, today it is about 1,000. A few decades ago Ayrshires, Guernseys, Jerseys and the Dairy Shorthorn, were major dairy breeds. Today they are slowly being lost from American farms. In 1880, Vermont's whole economy was based on the Merino sheep, now there are only two pure Vermont Merinos left.'

Similar societies are at work in Europe, trying to save what remains of their diverse agricultural heritage. In Chapter 9, breed details are given of those animals which are of special interest to the Rare Breeds Survival Trust and similar organisations in other countries.

9 The Breeds

The Rare Breeds Survival Trust lists all rare breeds of farm livestock under six category headings, according to their status at the last survey (1982). Further updated information can be obtained from the RBST.

Breeds of Cattle

Priority 1: Critical

1. IRISH MOILED

Description: Medium sized, polled. White back strip of varying width, differing amounts of white on other parts of the body. Adult cows 8 cwt, bulls 9 cwt.
Distribution: Counties, Antrim, Down and Donegal.
History: Cattle Society formed in 1926 to develop dual-purpose animals for economical production of milk and meat from poor land. Suitable for small scale farmers in Northern Ireland.
Present UK Population: Under 50 breeding females. Small groups can be seen at Croxteth Farm Park, Liverpool, and at Temple Newsome, Leeds.
Additional facts: Very rare indeed.

An Irish Moiled bull at Croxteth Park, Liverpool.

An Irish Moiled cow at Croxteth Park, Liverpool.

2. KERRY

Description: Small lean dairy type. Graceful horns on a fine light head. Very popular in Ireland in the nineteenth century and was then found in various colours. At the present time it is established as a black breed. Adult cows 7–7½ cwt, bulls 8 cwt.

Distribution: Mainly found in Irish Republic, sparse in UK, but small groups over many parts of the world.

History: Probably direct descendant of Celtic Black cattle and related to the Spanish Fighting bulls. Able to survive and produce a fair yield of quality milk from poor pasture.

Present UK Population: About 20 breeding females in small herds in different parts of the country.

Additional facts: Rare and specialised breed and has little commercial use at the present time.

3. GLOUCESTER

Description: Medium-sized dairy breed, usually dark mahogany in colour, with white strip extending from the middle of the back, over the tail, down the hindquarters and forward underneath the belly to the front legs. The horns are short.

Distribution: Mainly found in Gloucestershire and in some farm parks in other parts of the UK.

History: Possibly derived from the old (now extinct) Glamorgan breed of cattle and developed in Gloucestershire. Due to the small fat globules in its milk, it became popular in the production of Double Gloucester Cheese.

Present UK Population: Just under 100 females in small groups, usually owned by rare breed enthusiasts. Can be seen at the Cotswold Farm Park, Guiting Power.

Additional facts: The few representatives of the breed today have been developed as dual-purpose cattle. A breed society was formed in 1919 but ceased to function by 1945. Only two herds survived, one of which was dispersed in 1966 and included the one remaining pure bred bull. The remaining herd at Wickcourt in Gloucestershire was sold in 1972. Gloucester cattle were noted for their docility and tenants of the Duke of Beaufort were encouraged in the earlier years of the century to keep them, as they were never stampeded by the Beaufort Foxhounds. They have little commercial use at present.

A typical Gloucester cow from the Wick Court herd in Gloucestershire. The primitive pattern of a white back stripe extends down the hind quarters, along the belly to the forelegs.

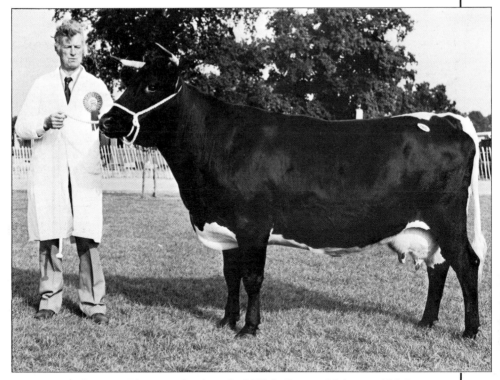

Joe Henson's champion Gloucester female at the RBST's Show and Sale 1984. This is one of the few animals of this breed still in existence.

4. SHETLAND CATTLE

Description: At the present time the breed is black and white and of medium size, rather short-legged, with a bulky body. The horns are short.

Distribution: Found mainly in their native Islands, but also in small groups in other parts of the UK. The largest herd is in Invernesshire, maintained by the Department of Agriculture for Scotland.

History: Believed to be of Scandinavian origin, it was developed as a crofter's cow, capable of producing some milk and a good beef calf from poor pasture. The cattle have been influenced by the introduction of breeds which were intended by absentee landlords to improve the local cattle. In 1870 White Shorthorn blood was used and in the 1920s Friesian bulls were put to the native animals, but due to the methods of farming and the climatic conditions natural selection took place. With the assistance of the RBST a carefully planned effort is now being made to strengthen the breed.

Present UK Population: About 100 breeding females in small herds and farm parks in the UK.

Additional facts: Seven Shetland Cattle were sent to the Falkland Isles by the RBST to help farmers to build up cattle stocks which were lost in the Falklands War of 1983. The conditions are similar to those found in the Shetland Isles: poor grazing and an exposed position.

A modern Shetland alongside a British Friesian cow. The latter breed has had some influence on the make-up of the present-day Shetland.

5. WHITE PARK CATTLE

Description: These cattle make up one of the most ancient breeds of British cattle and have now been developed into a true beef breed. On average, adult animals weigh 12½ cwt, have long horns, are white in colour with either black or red muzzle, eyelids, ears, teats and feet.

Distribution: The breed is kept throughout Britain in small groups, some in farm parks. Three of the old herds which were established in the thirteenth century are still in existence, the Dynevor Herd (formerly in Wales, now in Sussex), Cadzow Herd (Scottish Border) and the Chartley Herd (formerly in Staffordshire and Woburn Abbey, now in Norfolk). The Vaynol Herd, which was established in 1872 with animals from Scotland, is also still in being and now in the care of the RBST.

History: Two theories as to their origin are put forward. One suggests they existed in Britain many centuries before the Christian era, but the more acceptable theory is that they were brought to Britain by the Romans as sacrificial animals. They became feral during the Dark Ages and were later emparked and hunted by the Norman barons in the thirteenth century. Some herds, such as that which was kept at Vale Royal in Cheshire, and the Dynevor and Chartley Herds, were re-tamed possibly in the sixteenth century.

Present UK Population: Possibly between 150 and 200 females are kept in the different herds and farm park groups. The numbers are increasing every year.

Additional facts: White Park Cattle thrive in a wide variety of conditions, are capable of good growth rates and produce a quality lean carcase. The progeny are colour marked i.e. calves by a White Park bull crossed with any other breed of cattle are white with coloured points. The breed is extremely long-lived and individuals will still breed at sixteen years old.

The intelligence of the breed is well-known and is illustrated by the 'legend of the White Cow of Vale Royal'. Vale Royal in Cheshire, like so many country houses, was the scene of fierce fighting in the Civil War. The house was plundered by Cromwell's troops, who took amongst their loot the famous herd of white cattle with red ears. These animals were similar to cattle which were kept at Lyme Park, near Manchester. Cromwell's men drove the herd a great distance, but one old cow managed to escape and wandered back to her former home where she supplied the Royalist Cholmondeley family, who were on the verge of starvation, with milk for the duration of the Civil War. The family in gratitude rebuilt the famous herd of white cattle with red ears, domesticated them and used them for milk production. There they remained until the middle of the nineteenth century. What happened to the cattle is not recorded but they originated from animals which had been brought from Dore Abbey in Hereford by monks in the thirteenth century.

Another famous group which has survived many crises is the Chartley Herd. Emparked by the Norman Baron Ferrers of Chartley when he enclosed part of the Forest of Needwood in Staffordshire in the reign of Henry III, the cattle, together with deer and other animals, were prized for hunting. The Chartley cattle remained in the park until 1904, when a series of outbreaks of tuberculosis reduced the herd from around 50 animals to about 8. It was then decided that the only way to save the remaining animals would be to move them to a new environment. They were put up for auction with a reserve price of £1,000, the highest bid, however, was £100. Later the Duke of Bedford bought the herd privately to put in his park at Woburn Abbey.

A yearling White Park bull selected for breeding due to his rare blood line.

A White Park cow from Appleby Castle.

Tragedy struck again, however. During the journey by rail, the straw in the wagon caught fire and most of the animals were badly burned and as a result died. The only surviving bull broke his horns in the panic, the stubs turned septic and he died. The Duke, wishing to preserve what was left of the herd, put a Longhorn bull on to the remaining cows. The dominant White Park gene has, over the years, returned the cattle to their true type, the only Longhorn characteristic which remains is the length of the cattle's horns which are slightly longer than they were originally.

The Chartley Estate was sold in 1905, but a stuffed bull remained at the family home in Leicestershire. During the Second World War, the house was occupied by the military and the bull was sent to Leicester Museum where it stayed until 1964. The present Lord Ferrers, who lives at Hebden Hall in Norfolk, felt that the rightful place for the stuffed animal was in the possession of his family and, with the agreement of the museum, took it to his home. This act stirred his interest in the Chartley Herd, which had been associated with his ancestors for 700 years, and he decided to revive the herd.

In 1971 Lord Ferrers was offered the whole herd of 24 cows, so the cattle returned to their rightful owner. The herd now numbers about 70 head, of which 30 are breeding cows and 3 are stock bulls; the rest are followers. The best calves are reserved for breeding stock, and the others are sold fat. Breeding stock are sold annually at the RBST Annual Sale at the Royal Agricultural Centre at Stoneleigh. The beef is said to be very lean and full of old-fashioned flavour.

Lord Ferrers' aim is to preserve the cattle as a unique herd with their special characteristics of outstanding mothering quality, longevity and hardiness, and to this end no special breeding programme is used to transform the Chartley Herd into a modern breed.

Priority 2: Rare

BRITISH WHITE

Description: White cattle with coloured points, similar to those of White Park animals. The breed is hornless (polled) short-legged and medium sized. Mature cows weigh 11–12 cwt. Originally used as dual-purpose or dairy cattle, the beef characteristics are now being selected. British Whites are now mainly used as a suckling breed for beef production.

Distribution: Originally associated with mid-Lancashire and Cheshire, later with East Anglia, but now in many parts of the UK. The breed is becoming popular in America and Australia, due to the black teats which do not burn in the sun.

History: This breed can trace its history back for three hundred years. A herd of White polled cattle was kept in the Lord's Park at Whalley Abbey in Lancashire until 1697, when it was dispersed to Middleton Park in Lancashire and Gisburn Park in Yorkshire. The cattle from Middleton Park were moved to Gunton Park, near Norwich, in 1765, and at the beginning of the nineteenth century a herd was established at Blickling Hall. Later several herds were established in Norfolk with Blickling stock, and that county became the second home of the breed. Among other important herds was one maintained at Somerford Park in Cheshire. A breed society was formed in 1918. Until 1960 British Whites were kept mainly as dairy cows, but now their main role is in suckling herds for the production of beef. As herds gave up milk production and dispersed, the breed declined in numbers until the mid-1970s, and only the Heveringham Herd in Norfolk retained a considerable number of cattle. The Woodbastwick Herd was founded in 1840 and it played an important part in the survival of the breed.

Some authorities say the Whalley Abbey cattle were originally a selected, polled strain of the feral wild white cattle of that area and as such are closely related to the White Park cattle.

Present UK Population: Since 1980 the breed has become more popular and 36 herds are now known to exist. Probably up to 250 breeding females are retained at present.

Additional facts: MLC have officially performance tested British White bulls which have given a good account of themselves. They have a low back-fat measurement and good food conversion. The breed has been exported to North and South America and there is a thriving association in Iowa, USA, where beef of the breed commands a premium price.

British White bull 'Castleton Brendan' owned by C.O. Wright & Sons, Hampton Lucy, Warwickshire, said to be the heaviest and longest British White bull in Great Britain.

Priority 3: Endangered

1. LONGHORN

Description: A big dual-purpose breed, now mainly used for single suckled beef production. The long horns which grow in a variety of styles are its most distinctive feature. The body varies in colour from dark brown to light roan, but a white back stripe extends from the shoulders down the tail and along the underside.

Distribution: Originally the Craven district of Yorkshire, later they spread over the Midlands and to most counties of England.

History: Bakewell used the Longhorn in his improvement programme and selected for the beef-producing qualities, and for a short time they became Britain's most popular breed. Before Bakewell's time, the breed was a dairy or dual-purpose breed and almost every county in the North and Midlands had its own strain of Longhorn.

Present UK Population: The breed is increasing in numbers every year, possibly well over 500 breeding females at the present time. Can be seen in farm parks and are often kept in parks attached to stately homes, where they are both productive and ornamental.

Additional facts: In the seventeenth and early eighteenth centuries these cattle were used for draught purposes as well as supplying some milk and producing a beef calf. They now are specialist beef producers as the dairy strains have disappeared. Famous Longhorns of the past are Bakewell's 'Twopenny' and 'Shakespeare'.

A champion Longhorn bull from the Marquis of Cholmondeley's herd in Cheshire.

2. BELTED GALLOWAY

Description: Medium-sized polled breed, adult cows weigh about 8½ cwt. Galloways are usually black in colour, but dun and white are also to be seen. Belted black or dun animals have a white belt completely encircling the body between the shoulders and hips. Their long hair has a dense undercoat.

Distribution: Found in small numbers throughout the UK and other countries, particularly the USA and Canada, Australia, New Zealand, and South America.

History: The Galloway Cattle Society was founded in 1877, but it was not until 1921 that a Belted Galloway Society was formed and they became known as a separate breed. The belted, or sheeted, character has been known to exist for hundreds of years in cattle the world over, and is controlled by dominant genes. The Galloway's belt may have been produced by crossing with a Dutch breed, the Lakenvelder. The White Galloway, which has dark points like the White Park, became a separate breed in 1977.

Present UK Population: Under 1,000 breeding females in the UK. Some commercial herds in Scotland, used for the specific purpose of cross-breeding with White Shorthorn bulls to produce Blue-Grey suckler cows.

Additional facts: Belted Galloways have their own breed societies in the USA, New Zealand and Australia. They are also able to tolerate the heat of South America. The ancestors of the breed are the old Celtic cattle of South West Scotland.

A Belted Galloway cow with her bull calf.

3. DEXTER

Description: Dwarf cattle with short cannon bones. Black is the usual colour, but red and dun are also to be seen. A horned, dual-purpose breed, thought to have been

A full-grown Dexter cow with Bill Stephen, Farm Manager at Croxteth Park. The dwarf characteristic of the breed can be clearly seen.

selected from the Black Celtic cattle of Ireland, probably a dwarf strain of the Kerry breed. Mature cows weigh 6 cwt and measure about 40 inches at the shoulder.

Distribution: Found throughout the UK and small groups have been built up in several other countries, particularly South Africa.

History: The Breed was first recorded in Ireland during the eighteenth century and first imported to England in 1882. A Mr Dexter, who was agent to Lord Hawarden on his estate in County Kerry, is said to have selected dwarf cattle from the local type and developed the breed which bears his name.

Present UK Population: Possibly about 500 pure bred females. Can be seen in commercial single suckler herds and in some farm parks. Usually kept in small numbers on self-sufficiency holdings.

Additional facts: Originally a dairy breed, but now mainly dual-purpose. The breed temperament is not totally reliable and they occasionally produce deformed or bulldog (faced) calves.

Priority 4: Watching Brief

RED POLL

Description: Medium-sized dual-purpose breed. As the name suggests, red in colour and hornless (polled). White hairs are present on the switch of the tail and occasionally on the udder.

Distribution: In many parts of the world. Over most of the UK, more especially in the South Eastern Counties.

History: Originally a cross between the Suffolk Dun, which was a high-performance

A Red Poll cow. The breed is naturally hornless.

dairy cow, but is now extinct, and the hardier more beefy Red Norfolk breed. Originally very popular in its home territory, but lost ground to the Friesian breed.

Present UK Population: 2,000–3,000 pure and grading-up animals, mainly in the South and East of England.

Additional facts: Due to the scarcity of blood lines brought about by the drop in numbers in the middle of this century, Danish Red bulls have been imported and crossed with the breed to raise its productivity.

Priority 6: Feral

THE CHILLINGHAM HERD

Description: Very like the White Park Cattle, and with common ancestry to them. The horns are more lyre-shaped, sweeping outwards and upwards. The ears of Chillingham Cattle are always red, although the remaining points, the muzzle, hoofs and horn tips, are black. The herd has never been known to produce a coloured or partly coloured calf, unlike other White Park cattle. The animals are of medium size, smaller than they used to be, and this has been put down to their constant in-breeding.

Distribution: Chillingham Park, Alnwick, Northumberland, has been their home for many years, but a small reserve herd has now been set up in Scotland as an insurance against a catastrophe taking place at Chillingham.

History and Additional Facts: (Information supplied by the Dowager Countess of Tankerville, President of the Chillingham Wild Cattle Association Limited.) The Chillingham Wild Cattle, sole survivors of their species to remain pure and uncrossed with any domestic cattle, still roam in their natural surroundings over Chillingham Park, originally 1,100 acres but now reduced to 300. Their origin is uncertain, but the herd is thought to have been at Chillingham for the past 700 years. The ancestors of the herd roamed the great forest which extended from the North Sea coast to the Clyde estuary, and the successful capture of a number of these wild cattle in the thirteenth century would not only have eased the local food situation but would also have made it impossible for raiders to take such cattle. Being very wild and extremely fierce, they could not have been driven, like their domestic cousins.

In recent years blood samples have been taken from several of the wild cattle just prior to their deaths, and from them Dr J.G. Hall of the Edinburgh Animal Research Breeding Organisation has found the blood grouping to be unique among Western European cattle. This fact, therefore, adds more mystery to their origin.

For 700 years they have been in-breeding, yet old bones found in the park show that the only effect on the animals has been that they are now somewhat smaller than they used to be. Their remarkable survival is thought to be due to the fact that the fittest, strongest bull becomes 'King' and leader of the herd and during his 'reign' sires all the calves. When he is defeated by a younger contender, he retires from the herd for some time and becomes irritable and very dangerous to approach. During his temporary banishment he remains in sight of the herd which he later rejoins, but never again takes on the role of sire.

The bulls are seldom seriously injured in fighting and only on three occasions this century has one been killed in this way.

Chillingham cattle resting. One cow remains on watch and alert at all times.

Between the two World Wars, herd numbers remained fairly constant at between 35 to 40 animals. In January 1947 the herd was made up of 33 cattle. Then followed in Northumberland the most severe winter in living memory. The fourth blizzard struck the area in March, causing snowdrifts up to 40 ft (12 m) deep in Chillingham Park. Twenty of the herd died, leaving only 8 cows, 5 bulls and no young stock. During the next twelve months no calves were born and the famous herd faced extinction. In August 1948 a calf was born. Unfortunately, it turned out to be a bull, but gradually the herd began to replenish its numbers and ten years later, in 1958, stood at 24 animals, 11 of which were cows and 2 were female calves.

Photograph of a Chillingham bull at the end of the nineteenth century.

In 1967 foot and mouth disease came to within two miles of the Park. If the cattle had contacted the disease the Ministry of Agriculture would have been, by law, forced to carry out their slaughter policy. Once again catastrophe was averted. However, such good fortune could not be trusted to luck in the future and it was decided to set up a small reserve herd in Scotland.

The herd in 1984 numbered 8 bulls, 29 cows, 3 heifer (female) calves, and 4 bull calves.

The Chillingham cattle are completely wild and will kill a calf that has been handled by man. It is not possible to give them veterinary attention. Fortunately this would rarely be required as they seldom suffer from any disease. If a cow, especially an aged animal, gets into difficulties during calving, nothing can be done to assist her and both she and her calf are later found dead.

The cattle will not accept normal agricultural practice, and will refuse grain and other cattle food, but will eat a little meadow hay or straw if they are on the verge of starvation, such as was the case in the winter of 1947. In past decades when they roamed over a vast area they were able to search out winter food such as dead grass, leaves, tree bark and branches, but now, being confined to 300 acres, the food supply in winter is more restricted.

A Chillingham bull in 1985, showing how little the breed has changed in almost one hundred years.

The Chillingham Wild Cattle Association was formed in 1939 as a charitable organisation to take over the care and maintenance of the herd. At that time the then Lord Tankerville, whose family owned the cattle, became aware of the fact that the increasing costs of maintaining the animals was rising beyond his private means. When, in subsequent years, the Association proved capable of standing on its own feet financially, he arranged to bequeath the ownership of the herd to the Association. This took effect upon his death in 1971.

A further threat to the herd arose after the death of the 9th Earl of Tankerville in 1980, when it was decided to sell the Chillingham Estate. However, with the personal intervention of the Duke of Northumberland, the Park and its surrounding woodland was purchased by the Sir James Knott Charitable Trust which granted the Association a lease of the grazing rights for 999 years, thus ensuring for the foreseeable future the Wild Cattle of Chillingham.

In 1790 Thomas Bewick described the hunting of Chillingham bulls: 'On notice being given that a wild bull would be killed on a certain day, the inhabitants of the neighbourhood came mounted and armed with guns, etc, some times to the amount of an hundred horse and four or five hundred foot, who stood upon the walls or got into the trees, while the horsemen rode off the Bull from the rest of the herd, until he stood at bay; when a marksman dismounted and shot. At some of these huntings, twenty or thirty shots have been fired before he was subdued. On such occasions, the bleeding victim grew desparately furious from the smarting of his wounds, and the shouts of savage joy that were echoing from every side. But from the number of accidents that happened, this dangerous mode has been little practised of late years, the park-keeper alone generally shooting them with a rifle gun at one shot.'

One of the last great hunts is said to have taken place in 1826 when a bull was shot by Earl Clanwilliam. On October 15 1872 the Prince of Wales (later Edward VII) shot a bull, but instead of riding out the bull a quiet approach was made to the herd in a hay cart from which the shot was fired. Today animals which need to be culled are swiftly dispatched with a rifle.

Visiting the Chillingham Wild Cattle:

Chillingham Wild Cattle Association Limited,
Reg. Office: Estate House, Chillingham,
             Alnwick, Northumberland,
             Tel. Chatton 213.

The cattle can be seen in Chillingham Park, which lies 14 miles north of Alnwick, from 1st April to 31st October.

Breeds of Sheep

Priority 1: Critical

1. PORTLAND

Description: Small animal, weighing 70–75 lb. Both sexes are horned, with brown or tan faces and legs. New born lambs are foxy brown in colour, but change to grey or white during the first year. The wool is close, fine and short, coarse red fibres are found in the fleece, especially in the breech area. The face and legs are tan coloured and clean and the ears are small.

The breed produces high quality meat with a fine texture and excellent flavour. Usually only produces one lamb, but occasionally lambs out of season.

Distribution: Once concentrated in large numbers in the Dorset area. Now to be seen only in farm parks. The Portland is the rarest breed in Britain.

History: The first improved type of sheep resulting from man's selection of animals more suited to his needs, were the small, horned, tan-faced animals. The Portland, like the early Welsh Mountain sheep such as the Rhiw, belonged to this group. The latter breed is now thought to have died out, leaving only the Portland and, to a lesser extent, the Exmoor Horned and Dartmoor, as present-day representatives of this important group, which first took part in our ancestors' livestock breeding programme.

Present UK Population: Under 200 animals, which are mainly kept in farm parks or experimental stations.

Additional facts: The first Portland sheep are said to have swum ashore on the south coast of England, from sinking ships of the Spanish Armada. The fact that they have the same characteristics as Spanish sheep, such as breeding out of season and long, outward spiralling horns, adds credence to the story, but imported Spanish Merino blood may be a more simple explanation. The Dorset Horn sheep is derived from the crossing of Merino and Portland breeds. The Portland may find a role in future breeding policies, but has no commercial use at the present time.

A Portland ewe and ram.

2. LEICESTER LONGWOOL

Description: A large, hornless, white-faced breed, with a long, lustrous, curly fleece, which almost covers its legs and face.

Distribution: Yorkshire Wolds and North Humberside.

History: Once probably the most famous British breed of sheep. Used by Robert Bakewell in his quest for his ideal meat animal. The breed influenced the formation of many other breeds in the UK, Europe and elsewhere. The Border Leicester, Blue-faced Leicester, Cheviot, Lleyn and Wensleydale, as well as the Ile-de-France, are also made up of Leicester blood. The Dishley Leicesters used by Bakewell during the eighteenth century in his ram hiring schemes, probably hold the record for the most valuable farm livestock of all time.

Present UK Population: 300–400 animals at present. The breed has little role to fulfil.

Additional Facts: Is capable of producing heavyweight lambs on a diet of arable crops such as turnips.

Priority 2: Rare

1. COTSWOLD SHEEP

Description: Very similar to the Leicester and developed from the original importation of Roman Longwool. Further developed by crossing with Leicester rams.

Distribution: Found in small groups at farm parks and in its native Cotswold region of Gloucestershire.

History: The ancestors of the Cotswold breed made up the huge Roman flocks which ranged the sheep walks of that area and during this time and later, from the fourteenth century, played an important part in the English wool trade. During the eighteenth century a greatly improved animal was developed by the use of Robert Bakewell's improved Leicester rams.

Present UK Population: Probably about 300 breeding females. At present the breed is little used commercially.

Additional Facts: A large breed of sheep, once used for crossing with other breeds to produce big lambs, which were shorn before they were slaughtered. Cotswold sheep produced juicy, fat mutton, as opposed to early lamb.

2. LINCOLN LONGWOOL

Description: Very like the two previous breeds (Cotswold and Leicester). Large, polled sheep with long, lustrous wool.

Distribution: Largely restricted to its native Lincolnshire, but small groups in other parts of the UK. Large populations are found in South America and Eastern Europe.

History: Another breed developed originally from the Roman importation of longwools. Further improved by the use of Leicester rams. The breed produces lean lamb, but with a high percentage of bone.

Additional Facts: The breed was formed to live on the rich, marshy soil and tolerate Lincolnshire's cold winter months. Their long wool protected them from the biting easterly winds in fields which did not afford the protection of hedges. The high quality wool was well adapted to the worsted markets and technology of the early eighteenth century. The breed still has a flourishing export market.

Present UK Population: About 600 breeding females. The majority are found in Lincolnshire.

Description: Similar to the other longwool breeds. The wool is lustrous, long and open, but divides into uniform little knots or purls. The breed is large and polled. The skin of the face, legs and ears, is blue in colour.

Distribution: Mainly located in its native Yorkshire, but several flocks in other parts of the UK.

History: The breed was created in the nineteenth century by mating a Leicester ram to a Teesdale ewe. Wensleydale rams were traditionally mated with hill ewes to produce the famous Masham type of sheep.

Present UK Population: Possibly 300–400 breeding females.

Additional Facts: The dark pigmentation of the skin enables the breed to tolerate hot climates. The lambs of the breed are apt to lack vigour for the first twenty-four hours, but thereafter thrive. The breed is capable of producing a very lean carcase which has an excellent flavour. Ewes are long-lived and retain their good breeding quality.

A yearling Wensleydale ewe in full fleece. The breed is mainly located in its native Yorkshire. The dark pigmentation of the skin enables the breed to tolerate hot climates.

4. MANX LOGHTAN

Description: The Manx word for mouse-brown is 'loghtan'. This describes the consistent colour of this multi-horned breed. The horns in both sexes are variable, being usually one or two pairs, but six-horned or polled animals have been known. The tail is short. Adult ewes weigh about 90 lb.

Distribution: Possibly one third of all the pure-bred animals remain in their native Isle of Man. Nevertheless, due to the RBST's efforts, many small flocks have been established in the UK. Most farm parks have a group of these spectacular animals.

History: This very old breed is most likely to be of Scandinavian origin. Fragments of a woollen cloak found in a Viking grave have been carefully examined and found to be very similar to modern Loghtan wool. The breed has been little improved and still retains its primitive looks and habits.

Present UK Population: 300–400 animals, of which one third remain on the Isle of Man.

Additional Facts: Able to thrive on poor land, and are very hardy. Animals grow larger when moved to better land and are capable of producing a good, lean carcase. Soay blood is thought by some authorities to have been introduced to help to build up numbers, and in the last century Linton rams were introduced to the Island.

A Manx Loghtan ewe with twin lambs.

5. NORTH RONALDSAY

Description: A small breed which varies in colour from white through grey, brown and black, and many combinations of these. Adults weigh about 55 lb. The tail is short and the majority of animals are horned.

Distribution: The majority of animals are located on the Island of North Ronaldsay, but flocks are kept privately and in farm parks in other parts of the UK. The RBST has a large flock on its island of Linga Holm.

History: North Ronaldsay is the most northernly island in the Orkney group. The Gulf Stream alleviates the effects of the Atlantic gales, but the island being so exposed has a high moisture atmosphere. One effect of the gales is that beds of seaweed which grow in profusion in the relatively warm waters are torn up and thrown on to the beaches.

This seaweed provided the basis for a thriving iodine extraction industry. To prevent the farm livestock from soiling the valuable seaweed, a wall was built round the island to exclude the animals from the foreshore. As new, more viable industrial processes were developed, the iodine industry became unprofitable and the islanders returned to their original occupation of crofting. The sheep were, therefore, banished to the foreshore for nine months every year to live outside the sea wall. There, they developed the habit of eating seaweed, and now the North Ronaldsay sheep exist on it entirely, and have adapted very efficiently to this unusual diet. They are specific in their choice of seaweed, eating only kelp which is exposed mainly at low tide, rather than the wrack which lies further up the beach. The sheep, therefore, time their feeding according to the tides and need to satisfy their appetites at the appropriate time.

Present UK Population: There are believed to be about 2,500 breeding females on the native island and other islands. The administration on North Ronaldsay is the local Sheep Court, which consists of eleven farmers who are sworn in by a Justice of the Peace, and serve for a period of three years. They are responsible for maintaining the wall in good condition. The ownership of the sheep is determined by lug or ear marks which are recorded officially.

Additional Facts: Round the wall are nine pounds (pens) where the sheep can be gathered for ear marking, selection and castration. Breeding policy is based on the selection of strong ram lambs which are kept in the ratio of one to every twenty ewes. The breed produces under favourable conditions a good proportion of triplets.

The RBST own a neighbouring island and in 1974 a breeding unit of 160 animals were transferred to Linga Holm where they were allowed to run as a feral flock. Though there is a certain amount of predation of lambs by birds such as the Greater Black-backed Gull, the flock due to the mothering qualities rear an average 1.5 lambs per ewe each year.

As a result of the specialist diet to which the sheep have become adapted, North Ronaldsay animals when moved to a different environment suffer various problems, one of which is copper poisoning. This is said to be due to their ability on their native island to use what little copper they receive in their diet very efficiently. When they, therefore, receive a normal amount of copper it has the effect of a toxin and can lead to death from copper poisoning.

Priority 3: Endangered

1.   OXFORD DOWN SHEEP

Description: The heaviest of the Down breed of sheep, the Oxford is polled, with a dark face and legs. It has strong bones and thick fine wool of good length without any curls. Its legs appear short and its body barrel-like.

Distribution: The breed is most adaptable and flourishes everywhere, it is found in most countries of the world and widespread in the UK.

History: First developed in the middle of the nineteenth century by crossing short-woolled Hampshire or Southdown ewes with long-woolled Cotswold rams. The breed was developed to produce early maturing lambs, which yield a good carcase at 12–16 weeks old.

Present UK Population: Between 800 and 1,000 ewes which are spread all over the UK.

Additional Facts: Lambs sired by Oxford Down rams are early maturing and have a high growth rate. Due to this, combined with their ability to produce a substantial fleece of fine wool, the breed has been exported in large numbers to the USA, Portugal, Rumania and the USSR, and is increasingly in demand in Britain.

2. SHROPSHIRE

Description: Medium in size and polled. The breed has a soft, black face. The fleece is thick and typical of all shortwool breeds.

Distribution: Shropshire sheep, though found mainly in the Midlands and Welsh Border regions, thrive well almost anywhere.

History: The breed was derived from the local Shropshire and Staffordshire types, such as the Morfe Common and the Longmynd, with the infusion in the mid-

A Shropshire sheep at Shugborough Park Farm.

nineteenth century, of Southdown blood. Improved Leicester rams may also have been used.

Present UK Population: Up to 1,000 ewes are thought to exist at present, mainly in the Midlands.

Additional Facts: The breed is hardy and prolific. It was, in the early part of this century, exported to North America, East Africa and Australasia, where it is highly valued for its ability to produce quality lambs out of Merino type ewes.

3. WHITE-FACED WOODLAND

Description: One of the largest hill breeds of sheep, it has a white face and legs, with pinkish nostrils (an unusual characteristic in British hill sheep). Both sexes have horns, those of the ram being heavy and spiralled. The tail is very long.

Distribution: Mainly found in their Pennine home and the West Riding of Yorkshire, but flocks have been established in other hilly areas of the UK. Usually to be seen in farm parks.

History: Improved by the use of Merino rams on the local ewes, with the intention of improving the fleece quality of the Pennine breeds.

Present UK Population: Some 3,000 ewes are thought to be kept in commercial flocks, mainly in their native Pennines.

Additional Facts: It is said that during the eighteenth century the Duke of Devonshire, wishing to improve the type of sheep kept by his tenants at Chatsworth in Derbyshire, bought Merino rams from the Royal flocks at Kew. These he distributed around the estate and this cross-breeding brought about the White-faced Woodland sheep.

The breed has been crossed with other hill breeds to impart its size and vigour. If kept in lowland areas the breed gives a good account of itself. It is sometimes referred to as the Penistone breed.

4. HEBRIDEAN SHEEP

Description: A small, black sheep, in which both sexes are multi-horned. The tail is of medium length (sometimes referred to as half-tailed). Previously the breed was known as the St Kilda. Adults ewes weigh 80–85 lb.

Distribution: Flocks are scattered throughout the UK.

History: Originally, like the Manx Loghtan, of Scandinavian origin, but kept over the past centuries as parkland sheep to graze around stately homes.

Present UK Population: Possibly about 1,000 animals kept in farm parks and stately homes.

Additional Facts: The striking features of the breed are its colour, usually jet black, and its impressive horns, one pair of which usually curl downwards while the other pair grow almost upright, giving it the appearance of a devil. It has survived as an ornamental breed, but also produces a prime, lean lamb.

5. SOAY

Description: A deer-like, graceful animal, usually tan or a darker brown in colour with light shading under the belly and rump. Both sexes are usually horned, occasionally ewes are polled. Adults weigh about 50 lb. The tail is very short.

Distribution: Large feral populations on the archipelago of islands which make up the St Kilda group. Up to 100 years ago, they were known only on the island of Soay, from which they get their name. Also kept, like the previous breed (Hebridean), in parks attached to stately homes. Small numbers are in commercial use and kept for genetic conservation.

History: The Soay is the most primitive of all the breeds of sheep and is thought to have changed very little since it was brought to Britain by the first farmers, after the last Ice Age. Though feral in this country, it is regarded as a link between wild and domestic sheep. Its behaviour is unlike other breeds, in that it is most difficult to drive or herd with a sheepdog.

Present UK Population: Possibly 3,000 animals are thought to exist, over half of these being feral in their native islands.

Additional Facts: Soay sheep shed their short, hairy fleece naturally every year. They are hardy and very resistant to disease. Management is, therefore, very minimal. Groups of animals are kept by such companies as English China Clay Limited in Cornwall, for the reclamation of spoil tips. They require very little attention and are unlikely, due to their size, to trample and cause the tips to slide. They have proved to be ideal for re-establishing vegetative cover where it would otherwise be almost impossible or prohibitively expensive. Soay sheep produce a lean carcase at any age.

Priority 4: Watching Brief

1. SHETLAND

Description: The colour varies from white, through grey and black to light brown (moorit) in that order of dominance. The ewes are polled and the rams horned. The tail is short and appears to be flat.

Distribution: The majority is still located in the Shetland Isles, but due to the home weaving hobby which is popular at present, flocks have been set up in various parts of the UK.

History: Wool of the breed is world famous and forms the basis of the Shetland wool industry. Due to selection over a long period, the quality is the finest of any wool produced by British breeds. The breed is related to the other primitive breeds of the Scottish Islands, such as the North Ronaldsay.

Present UK Population: A large population exists on the Shetland Islands and over 1,000 animals are maintained in registered flocks in other parts of the UK.

Additional Facts: The famous 'Fair Isle' products are made from the wool of the Shetland breed. The ewes are good mothers, being hardy and thrifty and capable of producing a high lamb output from a relatively low intake of food.

2. RYELAND

Description: A medium-sized breed which is white-faced and hornless. It has a very symmetrical shape. The wool, which grows thickly and is closely set on the skin, covers almost every part of its body, leaving only the nose bare.

Distribution: Found widespread in Southern England.

History: The original Ryeland sheep was developed in the southern part of Herefordshire called Archenfield, on land which once grew a great quantity of rye, and was a cross between the native sheep of Hereford and possibly a Southdown ram.

In 1837 Youatt wrote of the Ryeland sheep as being one or our hardiest and most valuable breeds and being able to thrive on scanty fare.

Sir Joseph Banks, King George III's agricultural advisor, said the breed deserves 'a niche in the Temple of Famine'.

A Ryeland ram showing the typical dense wool.

The exceptional fineness of the original Hereford Ryeland wool has, to a great extent, been lost. So, too, has the exceptional quality of the meat, and rams are now used to produce quality lightweight lambs.

Present UK Population: Possibly up to 1,500 breeding females, found mainly from Hereford southwards.

Additional facts: The Ryeland was said to thrive on cold, damp soils and would survive where many other breeds would starve. The breed has been known as the Hereford, Archenfield and the 'Ross breed' and the earliest records of these sheep date back to the fourteenth century. In Elizabethan times Ryeland wool was said to be the 'Leominster ore' or the 'golden fleece' of the Leominster wool trade.

3. WILTSHIRE HORN

Description: The most remarkable feature of the breed is its lack of wool. Both sexes are horned, and weigh 160–170 lb.

Distribution: A number of flocks in parts of the UK, especially the Midlands and Anglesey. The breed, due to its thin covering of wool, tolerates hot climates and is less likely to suffer badly from maggot fly. Animals have been exported in small numbers to North and South America, parts of Africa including Zimbabwe, Australia where they have been the subject of some interest, and to Malaysia for crossing with native breeds.

History: Originally the breed belonged to the white-faced, horned sheep of South West England. The characteristic of losing their wool was selected for, and is unique among the improved British breeds.

Today the fleece consists of a thick, matted covering, consisting mainly of hair with very little wool present, and this is cast naturally every year. The breed was developed as a pure meat producer from the seventeenth century. George III used Wiltshire Horned sheep when he first stocked his estate at Kew; later he crossed them with Spanish Merino rams.

Wiltshire Horn rams from the internationally-renowned Gedwydd Flock, Anglesey.

The new polled strain of Wiltshire sheep owned by Mr Iolo Owen, Anglesey.

Priority 6: Feral

1.   BORERAY

Description: Adults weigh about 70 lb; both sexes are horned. The fleece is mainly grey or cream coloured, while the face and legs are black and white, or grey and white, short-tailed.

Distribution: Almost all the present population is in their native island of Boreray in the St Kilda group. The Animal Breeding Research Organisation near Edinburgh has a small experimental group.

History: Developed in the late nineteenth century from cross-breeding an early type of Scottish Blackface with a Hebridean variety of the old Scottish Dunface.

Present UK Population: 350–450 breeding females.

Additional Facts: The fleece is shed naturally in July. The breed has no commercial use.

Breeds of Pigs

Pigs were among the animals which survived the Ice Age in Britain, and the wild pig which the first farmers found in these islands when they arrived from Europe changed very little for hundreds of years. It was not until the late eighteenth century that imported animals from the Far East, mainly China, India and Siam, and the Neapolitan pig from Italy, brought about the development of a variety of different local breeds or strains. Importation of pigs into Britain in Roman times probably took place, but during the Dark Ages these died out, to leave the wild European pig and its half-tamed brethren as the sole representatives of the *Sus* species in Britain.

All pig breeds are, therefore, relatively new, unlike those of cattle and sheep.

Priority 1: Critical

1.   MIDDLE WHITE

Description: White in colour, with prick ears, a turned up nose and a dished face. Medium size.

Present UK Population and Distribution: Possibly up to 100 breeding females. Only a very few small herds survive. There are some Middle Whites in Japan.

History: Originated in the early 1900s as a pork breed, by crossing a Large White pig with a smaller white breed of Yorkshire. The short, flattened head characteristic is inherited from an ancestor which had been imported from Asia.

Additional Facts: Middle White boars were exported to Russia and Australia, as well as North America, in the early part of the twentieth century.

The breed is susceptible to diseases such as rhinitis, due to the impractical shape of its nose. It has very little role to play in today's pig industry.

A Middle White sow.

2.   LARGE BLACK

Description: A large, docile, all black breed, with lop ears. The skin and coat of fine hair are black.

Present UK Population and Distribution: Just over 100 pedigree females, found mainly in the southern half of Britain.

History: Breed society formed in 1899. One of the oldest pure breeds, having been developed from the black pigs of Devon and Cornwall, and the black pigs of Suffolk and Essex. Large Blacks were formerly fed to enormous weights to produce a very heavy, over-fat carcase.

Additional Facts: The breed was ideal for outdoor conditions, being a good mother, docile to handle and less liable to sunburn, due to the black pigmentation of the skin. It was also said to be hardy and thrifty. It was exported early in the century to the USA.

3. BERKSHIRE

Description: Small pig, black in colour, with white feet, white tail switch and a white mark on its partly turned-up snout. It has prick ears and a slightly dished face.

Present UK Population and Distribution: Under 100 pedigree females, found mainly in the Southern and Eastern Counties of England.

History: Originally the Berkshire was a large, reddish coloured pig, with black patches, and lop ears. The use of the imported Neapolitan strain which had been developed in Italy from the Asian pig, brought about a totally different type of Berkshire. This animal was more like the breed which we know today.

Additional Facts: The breed was developed as an early maturing pork breed which produced a quality lean carcase. Berkshires are well established in Australasia, from where stock has been imported back to the UK. The breed was exported to USA where it helped to provide the basis for the Poland-China breed.

A typical Berkshire sow, English strain.

4. TAMWORTH

Description: The only red coloured British breed. Medium-sized, with a long snout and prick ears.

A Tamworth sow.

Present UK Population and Distribution: Possibly 150 breeding females in Britain.

History: Was developed in Staffordshire and was most abundant in the Black Country. There are a number of theories which try to explain its red-gold colour, one being that a red boar from India was given to Sir Robert Peel and kept on his estate near Tamworth. This animal was used to mate with local sows and thus passed its colour on to its progeny. This resulted in the local strain of pigs all being red.

Additional Facts: The breed was popular as a producer of pork from waste products. It was hardy and adaptable to a wide range of conditions. The breed became popular in USA, Australia, and parts of Asia. Some years ago, when numbers in Britain fell very low, new blood lines were imported back from Australia to strengthen the breed.

Being a hardy breed of animal, which contained many of the instincts of its primitive ancestors, Tamworth sows were crossed with a wild boar by Joe Henson, owner of the Cotswold Farm Park, to take part in an historical project for television. The resultant cross produced what are now known as 'Iron Age' pigs, which are described by their developer as 'Henson's Hash-ups'.

5.   BRITISH LOP

Description: One of the biggest breeds of pigs. White in colour, with lop ears. Almost the same shape as the Large Black.

Present UK Population and Distribution: 150 breeding females, found mainly in the South West corner of England.

History: The breed contains blood from some of the oldest white strains of British pigs, such as the Cornwall White, Devon Lop and the White pigs of Wales and Northern Ireland.

Additonal Facts: Possibly modern Welsh, Landrace and Large White blood has been introduced to the breed. Although it is a multi-purpose pig, it has not spread far from its native area. A good outdoor breed of pig.

Priority 2: Rare

GLOUCESTER OLD SPOT

Description: A large, lop-eared pig, mainly white in colour, with one or two black spots. Originally it had a more liberal covering of spots.

Present UK Population and Distribution: Found mainly in the Berkeley Vale region, but herds have been established in other parts of Britain. Between 200 and 300 registered sows, with probably many more unregistered.

History: A breed society was formed in 1914. The Gloucester Old Spot originated in the Severn Valley where it was used to convert by-products, such as whey from cheese-making, and fallen apples from the cider apple orchards, into top quality bacon and pork for the London market.

Originally the black spots were far more numerous and were said by some to have been caused by bruising when the apples fell on to the pig's back. Selection has taken place and the modern Gloucester Old Spot has been developed by breeders such as Mr George Styles of Ribbesford, Worcestershire, to have only one or two spots and to be more suited to modern requirements.

The breed is dual-purpose, hardy and prolific. It has excellent mothering qualities and is a good forager.

Priority 3: Endangered

BRITISH SADDLEBACK

Description: A large breed of pig. Lop-eared, black animals, with a white belt which varies in width, running round the body from the shoulder backwards.

Present UK Population and Distribution: Found throughout the UK. Up to 300 registered females, but probably many more non-pedigree animals.

History: British Saddlebacks resulted from the amalgamation a few years ago of the Essex breed from East Anglia, and the Wessex from South West England.

Additional Facts: Once the ideal sow for outdoor pig production. When crossed with a Large White or Landrace boar, the offspring are mostly white with a few blue patches, and possess hybrid vigour. The herd is hardy, docile and prolific, and has been used to some extent in the production of many of the modern commercial hybrids which are fashionable at the moment. The New Hampshire pig of America is said to contain a high proportion of Saddleback blood.

A British Saddleback sow and litter.

Breeds of Goats

Priority 1: Critical

BAGOT

Description: The preferred colouring is a black head, neck and shoulders, with the rest of the body white, but this is rarely achieved. The hair is long, and both sexes are horned.

Present UK Population and Distribution: Possibly about 50 breeding females. Kept in zoos, farm parks and private collections.

History: Originally brought to Britain from the Rhone Valley in Switzerland by the Crusaders. A semi-feral herd which lived in Bagot Park, Staffordshire, accounted for most of the breed until 1976. When the park was sold the remaining goats were donated by Nancy, Lady Bagot, to the RBST.

Additional Facts: The RBST had proposed to import some Schwarzhal goats, which are thought to have common ancestry with Bagots, from Switzerland, but this plan has been delayed due to disease problems.

One theory about the origin of Bagot goats is that the breed may be a selected strain of feral British stock. This supposition seems to be born out by the evidence that natural selection favours Bagot-type genes. One example is the Bagot goats in the Rhinog Hills of Gwynedd. Five Bagots were released on to the hills in the 1950s. Now, through natural selection, over a third of the hundred plus feral goats are of this type.

Priority 2: Rare

GOLDEN GUERNSEY

Description: A medium-sized goat, with short or long golden hair on the body, face and legs. The skin is also golden. The ears are large and have a distinctive upward and outward turn at the tips. The breed is usually horned.

Present UK Population and Distribution: Probably 350 breeding females to be found throughout the UK.

History: Like the dairy cattle which share the same name and golden colour, the Golden Guernsey goats originate from the Channel Island.

Additional Facts: Until the late nineteenth century most types of European goats were very mixed, but certain characteristics were beginning to be selected for. Whether it was colour or horn shape, selection started to show special features in different types of goat. The Golden Guernsey may have developed at this time from the Golden Saanen (or Gessenay as it was known in Europe), with a little Alpine and Maltese blood. Or it may have been selected for hundreds of years on its island home, having been first taken there by monks in the ninth century A.D.

During the Second World War, the breed escaped extinction solely due to the efforts of one enthusiast, Miss Milbourne, who is said to have hidden some animals in caves to prevent their being killed during the German occupation of Guernsey.

A Golden Guernsey milking female.

Breeds of Horses and Ponies

Priority 1: Critical

EXMOOR PONY

The ancient origin and unique ancestry of the Exmoor pony are reasons in themselves why the breed should be preserved. The story of these animals seems to begin when our early ancestors, the neolithic people from Europe, first reached the shores of Britain. The ancestors of the pony possibly arrived at the same time, or may even have predated them, having been survivors of the Ice Age, who inhabited the British Isles before the formation of the English Channel.

Ponies have been actively bred on Exmoor for many hundreds of years, but the descent of the present Exmoor Pony can be traced from 1820. In that year Sir Thomas Acland sold his Exmoor estate and took twenty ponies to found a herd at his new home, Winsford Hill.

Over the years, due to land improvement, agricultural depression and times of war, the breed was 'improved' with the use of other breeds of stallion. Only the Acland herd retained the true Exmoor type.

In 1952 the Acland herd, which numbered about a dozen animals, was reorganised and the Exmoor Pony Society produced a new stud book. Until 1961 this remained

Exmoor pony mares on Winsford Hill, Exmoor.

open to good mares of the Exmoor type, but since then has only accepted ponies from fully registered parents.

The breed is easily recognised by its primitive equine colour of dun, with an overcoat of dark brown, and its mealy-coloured muzzle and eye surround, the latter point exaggerates the 'toad-eye' which is a distinctive characteristic of the breed.

Mares stand 12.2 hands, stallions 12.3 hands. On its native moor, it is capable of living without housing on a diet high in roughage. Its extreme hardiness being due to natural selection over many years.

Priority 2: Rare

1. SUFFOLK

Today's Suffolks trace their ancestry to one eighteenth-century stallion. The breed is shorter than the Shire and the Clydesdale, but heavier, weighing up to 2,400 lb. The legs are short and hairless and give the impression of being too small for the rather heavy, muscular body. Due to this rather comic appearance, it was often referred to as the 'Suffolk Punch'.

Suffolks are always chestnut in colour, although different shades do occasionally occur. As a breed they were popular in America and the Colonies for crossing with native mares.

2. DALES PONY

Like the Fell pony from the western Pennines, the Dales pony from east of the Pennine watershed, is an animal of the uplands. Both were developed as pack horses capable of carrying heavy loads over the rough terrain.

They were also used as light draught animals and for riding, on the hill farms of thair native area.

The accepted height is 14.0–14.2 hands and the colour is usually black, sometimes with a white star and, in the Dales pony, one or two white feet.

Though their commercial use on farms has all but ceased, they are now being used as riding horses, especially due to their sure-footedness for pony-trekking.

Priority 3: Endangered

CLYDESDALE

Smaller than the Suffolk, being up to 17.0 hands high, the average being about 16.2 hands. The colour varies from brown and black to the less common grey. The coat often has a dappled appearance and the legs are lightly feathered (a certain amount of long hair grows below the hock and knee and covers the hooves).

The ancestors of the breed are the sturdy ponies of Southern Scotland, which were mated to Black Flanders and English Shire stallions. The Clydesdale Horse

Society was formed in 1877 and selection, together with the occasional dash of Shire blood, has taken place ever since.

Priority 4: Watching Brief

CLEVELAND BAY

A large, bay, riding horse, originating in Yorkshire. In medieval times a race of horses existed around the horse breeding district of Cleveland in the North of England and these were said to be clean-legged and bay in colour. First used as riding horses and later for coaching work for such runs as the Edinburgh to London route, the Cleveland was fast walking, sure-footed and had tremendous stamina.

Records show that for the last two centuries the breed has been kept pure, with no thoroughbred or carthorse blood having been added.

The Cleveland Bay Horse Society was formed in 1883 and the breed prospered, but in the early part of this century many of the best animals were exported to America. As its work as a coaching horse declined, so did its numbers.

The breed is still exported all over the world, to be used as an improver of native stock, Japan being one such country to buy Cleveland stallions.

In Britain it is used to sire both hunt and event horses.

A Cleveland Bay mare at Croxteth Park, Liverpool.

Unclassified Breeds

1.   JACOB SHEEP

Genesis 30, Verse 32 (1599 Edition).
Laban's shepherd, Jacob said:

> *'I will pass through all thy flocks this day*
> *and separate from them all the sheepe with*
> *little spots and great spots and all the*
> *blacke lambs among the sheepe. . . .'*

This passage gives birth to the legend about this ancient breed. Jacob was the son of Isaac and the unpaid shepherd of his greedy father-in-law, Laban the Syrian. For the love of his wife, Rachel, Jacob worked thus for fourteen years. It was then agreed that all the spotted sheep should belong to him in lieu of wages. By the application of early breeding techniques, he used a spotted ram and acquired the major portion of the flocks.

A Jacob lamb.

In the first instance the breed came from Palestine, via North Africa to Spain, with a helping hand from the Phoenicians, to eventually be wrecked in ships of the Spanish Armada and to arrive on the coast of the British Isles.

George Lucy is said to have imported Spotted Sheep from Portugal in 1756 to his home at Charlecote Park, Warwick. Nearly 200 years previously, Sir Thomas Lucy is said to have prosecuted William Shakespeare for game stealing from Charlecote Park, and this may give a clue as to the reason why Jacob sheep, or spotted sheep, are sometimes referred to as 'Shakespeare's Sheep'.

Experts do not readily agree with these legends, but say that the Jacob is related to the multi-horned sheep of the Northern short-tailed group. On comparison with, say, the Hebridean (formerly the St Kilda) some differences are obvious. It is, therefore, left for the individual to decide which theory to accept.

The Jacob is a medium-sized, multi-horned, white sheep with black or dark brown spots or patches. In 1969 it was dismissed as being just an adornment for country parks and stately homes. Today it is a popular breed, used commercially for wool and fat lamb production, and produces a first-class lean carcase when crossed with other breeds. Cross-bred lambs are usually black in colour. It has been exported to other countries and has proved to be a great attraction at farm parks.

2. NORTHERN DAIRY SHORTHORN

During the mid-sixteenth century, in the North East of England, a superior race of Shorthorn cattle was beginning to make its appearance. The cattle were of Dutch descent, but had a discernible admixture of Scandinavian blood. Two hundred years later, two similar types of high yielding dairy cows sprang to prominence, one in the Tees valley, which was known as the 'Teeswater' and another in a different part of Yorkshire, the 'Holderness'.

Robert Bakewell's pupils, the Colling brothers, took the Teeswater cattle and, using the well-tried methods which they had learned, developed the Shorthorn breed.

Later development by other noted breeders brought about the division of the breed into the Beef Shorthorn, Dairy Shorthorn, and Lincoln Red Shorthorn. In Australia the Illawarra Shorthorn maintained many of the characteristics of the old type.

In the dales of the Pennines, another strain was being developed. The Northern Dairy Shorthorn, being an animal of a harsh upland habitat, was hardy and thrifty, capable of producing good yields of milk and meat. The cattle were not pedigree, but were kept pure by the use of good Shorthorn sires. The breeders of these cattle were not influenced by fashion or trends, and refused to allow their animals to be spoiled by the publicity-conscious Breed Society.

During the Second World War, the urge to increase milk production quickly led to much cross-breeding of cattle and threatened to deplete the stocks of the specialist Pennine strain.

Several breeders with foresight decided in 1944 to set up the Northern Dairy Shorthorn Breeders' Society, with a view to preserving the purity of the breed.

Thus the Northern Dairy Shorthorn, which was basically a direct offshoot of the

A Northern Dairy Shorthorn.

early Teeswater cow with perhaps a little Ayrshire blood, became established as a pure breed.

The most popular colour is light roan, but red, white, and mixtures and shades of all three are found.

In 1969 the separate *Herd Book* ceased and the cattle were included in the *Coates Herd Book* with an 'ND' prefix. The total is about 200 cows and not more than 6 bulls.

The true Northern Dairy Shorthorn is, therefore, possibly eligible to be classed as a Priority 2 Rare Breed.

Appendix I
Artificial Insemination
of Farm Livestock

The practice of artificial insemination is very old, having been first used in the fourteenth century by Arab horse breeders. Sponges, impregnated with semen, were used as a means of fertilising mares.

In Italy, as long ago as 1780, the same method was used for dog breeding, and at the close of the nineteenth century artificial insemination was used on mares in Britain.

In 1909, a scientist name Ivanoff established a laboratory in Russia for the purpose of improving existing techniques. His particular interest was the use of AI as a means of controlling the spread of disease in livestock. By 1938, over one million cattle and fifteen million sheep had been inseminated in the USSR, and it was here where all the basic work was carried out.

By 1936 the Danish government was showing practical interest in AI and within eleven years had 100 cooperative breeding stations inseminating over half a million cattle annually.

The USA began to use the practice commercially in the following year (1937) and by 1942 the UK had done likewise. At the end of 1950, close on 100 centres and sub-centres were in operation in Britain, used by some 60,000 farmers.

In 1961 over one and a half million cows were inseminated in the UK from the Milk Marketing Board's Artificial Insemination Centres.

Today two million cattle are inseminated annually by the semen from over 39 breeds of bulls.

The Advantages of AI

A bull produces 50 to 100 times as much semen as is required for natural fertilisation. Therefore it is possible for a good animal to produce many more calves using the artificial technique. The semen is diluted and can be stored for many years when frozen.

The spread of disease is considerably reduced, as previously one bull may have been used by many farmers and disease contacted in one herd would have been rapidly spread to another.

Farmers with a small number of cattle have the opportunity to use semen from the best bulls in the UK, and are also able to use different breeds, for example dairy or beef, depending on the needs at the time.

A bull can be progeny tested and not used on a large scale until he has proved his worth.

Semen from rare breeds can be stored and used on almost any farm in the British Isles. A White Park bull, for example, can be standing at the National Agricultural Centre, yet is able to produce calves in herds from Land's End to John O'Groats.

AI in Pig Breeding

Several centres began to provide a pig AI service as a side-line to cattle AI during the middle of this century. In 1962 the Pig Industry Development Authority (PIDA) took an interest in the practice and two years later (1964) established a Pig AI Centre at Thorpe Willerby, Yorkshire. By 1965 a delivery service had been set up, whereby fresh semen was sent by express train to most parts of the country.

Difficulty has been experienced with the storage of pig semen, as once frozen it loses a great deal of its fertility.

Fresh semen gives an 80 per cent success, whereas with frozen only 40 per cent is attainable.

Appendix II
Rare Breeds and Education

The most famous rare breed collection is the internationally renowned *Cotswold Farm Park* at Guiting Power, Cheltenham. Owned and managed by Joe Henson, the unit contains the most comprehensive collection of rare breeds of British farm animals on display to the public in the country. It acts as an important 'shop window' for the Rare Breeds Survival Trust, of which Joe Henson was the first chairman.

Twenty-five acres of the twelve hundred acre Benborough Farm was originally set aside to accommodate the farm park, which became Approved Rare Breeds Centre No. 1. Set in one of the loveliest parts of England, the area is left in its natural state except for the addition of a few tree shelter belts which are intended to afford the animals some protection from the elements.

An indoor study centre, containing audio visual programmes with linked tapes and slides, and an education centre have been set up due to the volume of school parties and the demand for educational facilities. A farm trail is also laid out for visitors interested in modern husbandry and seasonal activities such as harvesting and sheep shearing can be seen at the appropriate time of year.

The National Trust's *Wimpole Estate*, eight miles southwest of Cambridge, has a comprehensive and unique collection of rare breeds of farm animals. The estate, once owned by Philip Yorke, third Earl of Hardwick, was bequeathed to the Trust in 1976 by Mrs George Bambridge, the daughter of Rudyard Kipling. Research into the records of the estate by the Trust, established the historic importance of the Home Farm and, with the help of the Historic Buildings Council, work began on restoring the eighteenth-century model farm. The historic parkland around the hall cannot be ploughed so the home farm has been managed as a productive livestock unit.

From the start the National Trust has worked closely with the Rare Breeds Survival Trust, and livestock that may have been seen during the 200-year span of farming at Wimpole form an integral part of the argricultural policy. Apart from a commercial flock of pure-bred Clun Forest sheep and some cross-bred suckler cows, all the breeds on the farm are either classified as rare, or have recently been taken off the list.

Mr Michael Rosenberg, Hon. Director of the Rare Breeds Survival Trust, generously gave the nucleus of the herd of Longhorn and White Park cattle, which now numbers some 60 head, to the National Trust. Also to be seen are British White cattle, the latter from the well-known Hevingham of Miss Birkbeck. Nine breeds of sheep are present, but the main interest is in the Leicester Longwool and Portland breeds. Other primitive sheep, such as Soay, are to be seen in the paddock. Six breeds of goats, including the rare Bagots, are exhibited in the collection and some are milked both by hand and by machine. A Tamworth pig is hopefully the forerunner of more rare breeds of pigs.

Together with the historic building and other tame animals such as donkeys, Shetland ponies, rabbits and guinea pigs, Wimpole Home Farm provides an enjoyable and educational day out for all visitors.

Many city and county councils, as well as individual schools, have developed or are in the process of developing rare breed collections. As zoos become less popular with the public, due probably to the excellent quality of television and film documentaries which show animals in their native habitat, rare breeds of domestic livestock are beoming a major attraction.

Schools have found it possible to use rare breeds of farm animals in an educational context. Not only can they be the subject of history and geography lessons, but also used for illustrating conservation generally and as a means of enabling town children to understand more about the ways of the countryside. On a more scientific level, the study of breeding techniques and the resultant inherited characteristics are equally well demonstrated by these animals. Not only do rare breeds provide subjects for such educational projects, but they are also a source of pleasure and interest to people of all ages.

A fine collection of rare breeds of domestic livestock can be seen at *Graves Park, Sheffield*. Owned and managed by the City of Sheffield Recreation Department, the farm animal unit has been developed since 1976 on twenty-two acres of agricultural land adjoining an existing mansion. The planning of the unit and its layout had the general public in mind, but now schools are finding it a valuable educational experience. The explanatory booklet provided by the Recreation Department states that since the beginning of the twentieth century more than twenty breeds of domestic livestock in the British Isles have become extinct. Changes in fashion and consumer requirements, combined with the pressures of modern intensive farming, have contributed to the decline and extinction of these breeds. Only recently have conservationists, farmers and scientists begun to realise the drastic loss of this genetic material, which once gone is lost forever. The booklet emphasises that rare farm animals are just as much a part of our national history as ancient buildings, and the following question is posed; can you imagine the countryside without horses, cattle and sheep? The animals in the park could well represent the past as well as the future of agriculture.

Merseyside County Council has a unit at *Croxteth Park, Liverpool*. Here again rare breeds of domestic livestock are exhibited. The general public, which normally has very little contact with animals, being mainly from within the city bounds, is able to experience livestock at close quarters. Explanatory notices give visitors information about the animals as well as helping them to understand Britain's farming progress over the years.

Croxteth Park is conducting a number of projects within the farm park. The only nucleus of Irish Moiled cattle in England is owned by the Merseyside Council. It is the Irish connection with the city of Liverpool which has prompted this interest in

Highland cows—an attractive addition to a parkland setting.

The only breeding group of Irish Moiled cattle in England, Croxteth Park, 1985.

the breed, whose numbers have been reduced in the last few years to danger level. The council hope that through a careful breeding policy enough cattle will be maintained to prevent the breed from being lost for all time.

The unit is also in the process of rebuilding the Lincolnshire Curly Coated breed of pigs (see page 89), as the last pure representative of these animals died in 1972. The Chester White, an American breed, was developed from the Lincolnshire Curly Coated in the middle of the last century and the Croxteth experiment will, it is hoped, due to careful selection of Chester Whites, result in the rebirth of a typical Lincolnshire Curly Coated animal.

Dean School Gloucester Old Spot taking part in Harvest Festival service.

Shugborough Park Farm was built in 1805 as the home farm to the Shugborough estate and now houses the Agricultural History Department of the Staffordshire county museum. Its farm buildings are being slowly restored to their original condition. The livestock are all breeds which originated from Staffordshire and have been used at Shugborough during the last two hundred years. The aim of the park farm is to illustrate the agricultural history of the county through a living exhibit. During the eighteenth century a herd of White Park cattle was kept, later to be replaced by improved Longhorn, Devon and Alderney cattle.

Southdown sheep were used to cross with black-faced ewes, and New Leicester and Merino rams were also to be found on the estate in the nineteenth century. In 1814 there were 1,300 sheep on the estate. Up to 70 pigs were kept in a hoggery in the centre of the yard and fed on dairy waste products. In 1794 the Staffordshire hog was described as large, slouch-eared and being whole-coloured or with black spots. Among breeds of farm livestock to be seen at the Park Farm today are Tamworth and Gloucester Old Spot pigs, Longhorn, Shorthorn and White Park cattle, Shropshire and Southdown sheep, and Bagot goats.

146

Another well-known collection which until 1981 was based in West Wales can now be seen at the National Trust's *Parke Estate* at Bovey Tracey in Devon. Parke Rare Breeds collection, which is owned by Mr and Mrs Timothy Ash, opened to the public at its new home in 1983 and provides visitors with the opportunity to see breeds of cattle, pigs, sheep and ponies which are either indigenous to Britain or were imported more than 100 years ago. Many breeds of goats, ducks, poultry and rabbits are also displayed. Tame young animals are exhibited in a pets' corner where visitors can experience close contact, being able to stroke and touch the animals. The National Trust and the Dartmoor Park Authority have combined to open Parke Estate to the public.

Also well worth a visit is *Temple Newsome* owned by Leeds City Council. It is an historic estate of over 1000 acres situated on the eastern outskirts of the city. The leisure services are in the process of developing what promises to be one of the largest rare breed units in Britain. Economically viable herds of ten rare breeds of cattle, ranging from White Park to Dexters, are maintained in groups of over 20 individuals of each breed plus their progeny. These herds of historically interesting animals make up not only attractive exhibits but also commercially sound units. One of the most recent introductions is a group of Moiled Cattle from Northern Ireland. This is only the second herd of such animals on mainland Britain. Sheep too are represented with a flock of over 100 breeding ewes made up of the Wensleydale, Black Welsh Mountain, Jacob and Hebridean breeds. Tamworth, Middle White and Gloucester Old Spot pigs are bred pure. The best of the surplus animals are disposed of annually as breeding stock and the remainder sold through commercial outlets. Visitors can see all aspects of farming at Temple Newsome.

Schools have found that rare breeds give an added dimension and incentive to pupils studying subjects such as natural history and biology, as well as history and geography. One such which has used rare breeds of domestic livestock on their school farm, is *Deane School, Bolton*. Under the guidance of Mr F. Tyldesley, Head of Rural Science, Deane School pupils have built and stocked a farm which, it is hoped, will soon become a financially viable unit. Pupils who would have little chance otherwise to mix with farm livestock at close quarters learn to take charge of the school's cattle, pigs, sheep and goats.

A place now seems to be assured for many of the less well-known breeds of domestic livestock and some are finding an increasing role to play in modern agriculture. The decline in numbers of some of the breeds will nevertheless need to be monitored consistently, and although the Rare Breeds Survival Trust plays a major role in this work it is only with the help and support of all people who value Britain's livestock heritage that the Trust is able to carry out its work of saving these interesting, ancient and valuable breeds.

Appendix III
Useful Addresses

Rare Breeds Survival Trust
4th Street
National Agricultural Centre
Stoneleigh Park
Kenilworth
Warwickshire CV8 2LG

The American Minor Breeds Conservancy
Box 225
Hardwick
Massachusetts 01037

Livestock Breed societies

Cattle

Belted Galloway Cattle Society
49 Tylers Acre Avenue, Edinburgh.

British White Cattle Society
89 Repton Rd, Hartshorne, Burton-on-Trent.

Chillingham Wild Cattle Assoc. Ltd.
Estate Office, Chillingham, Alnwick.

Dexter Cattle Society
Seckington Lane, Newton Regis,
Tamworth, Staffs.

Gloucestershire Cattle Society
1 Barton Cottages, Guiting Power,
Cheltenham, Glos.

Irish Moiled Cattle Society
Laurel Bank, Saintfield, Co. Down,
N. Ireland.

Kerry Cattle Society of England
East Johnston, Bish Mill, South Molton,
Devon.

Longhorn Cattle Society
1 Barton Cottages, Guiting Power,
Cheltenham, Glos.

Red Poll Cattle Society
6 Church Street, Woodbridge, Suffolk.

Shetland Cattle Herd Book Society
Parkhall, Bixter, Shetland.

White Park Cattle Society

Sheep

Black Welsh Mountain Sheep Breeders
Assoc.
Brierley House, Summer Lane, Combe
Down, Bath.

Cotswold Sheep Society
1 Barton Cottages, Guiting Power,
Cheltenham, Glos.

Leicester Longwool Sheep Breeders' Assoc.
Street Farm, Loftus, Saltburn, Cleveland.

Lleyn Sheep Society
Tyn Rhos, Llangybi, Pwllheli, Gwynedd.

Lincoln Longwool Sheep Breeders' Assoc.
Lincolnshire Showground, Grange de
Lings, Lincoln, LN2 2NA.

Oxford Down Sheep Breeders' Assoc.
Guardswell Cottage, Inchture, Perth.

Ryeland Flock Book Society
Brick Kiln Cottage, Portway, Burghill,
Hereford.

Shropshire Sheep Breeders' Assoc.
The Cattle Market, Worcs.

Southdown Sheep Society
41 Commercial Rd, Bedford.
Bedford.

Teeswater Sheep Breeders' Assoc.
Runley Bridge, Settle, North Yorks.

Torddu: Badger-Faced
Welsh Mountain Sheep Society
Cwnllechwedd Fawr, Llandrindod Wells,
Powys.

Wensleydale Longwool Sheep
Breeders' Assoc.
Bryn goleu Farm, Cornist Lane, Flint,
Clwyd.

Wiltshire Horn Sheep Society
Ham Farm, Ham Hall, Holcombe, Bath.

Goats

British Goat Society
Rougham, Bury St. Edmunds, Suffolk.

English Goat Breeders' Assoc.
Oak Cottage, Darsham, Saxmundham,
Suffolk.

Golden Guernsey Goat Club
East Johnstone, Bish Mill, South Molton,
N. Devon.

Pigs

British Lop Society
Trewelland, Liskeard, Cornwall

National Pig Breeders' Assoc.
7 Rickmansworth Road, Watford, Herts.

Horses

Cleveland Bay Horse Society
York Livestock Centre, Murton, Yorks.

Clydesdale Horse Society
26 Argyll Terrace, Dunblane, Perthshire.

Dales Pony Society
Ivy House Farm, Hilton, Yarm, Yorks.

Eriskay Pony Society
114 Braid Road, Edinburgh.

Exmoor Pony Society
Glen Fern, Waddiscombe, Dulverton,
Somerset.

Fell Pony Society
19 Dragley Beck, Ulverston, Cumbria.

The British Spotted Pony Soc.
Wantsley Farm, Broadwindsor, Beaminster,
Dorset.

Shire Horse Society
East of England Showground,
Peterborough, PE2 OXE.

Suffolk Horse Society
6 Church Street, Woodbridge, Suffolk.

Poultry

The Poultry Club of Great Britain
Cliveden, Sandy Banks, Chipping, Preston.

The Rare Breeds (Poultry) Soc.
8 St. Thomas' Road, Great Glen, Leics.

The British Waterfowl Assoc.
Five Gables, Evenley, Brackley, Northants.

*Enquiries about other breeds should be referred
to the RBST.*

Appendix IV
Minor Breeds In North America

Following is a list of minor breeds of livestock that have come to the attention of the American Minor Breeds Conservancy. Since AMBC has not completed its survey of minor breeds of livestock in North America, this list should not be considered comprehensive.

The breeds listed here are not necessarily comparable regarding their endangered status. Some minor breeds are being promoted heavily and recently have made rapid gain in numbers (e.g. Texas Longhorn cattle). Some breeds are considered minor because their numbers are declining rapidly (e.g. Ayrshire cattle). Some breeds do have active breed associations, but are still rare (e.g. Cotswold sheep). Other breeds seem to be quite numerous but have no active association (e.g. Lineback cattle). Still others have no breed association and are clearly endangered. We have included the names and addresses of breed associations where we have that information.

Cattle

Ayrshire
Belted Galloway
British White
Canadienne
Criollo
Dexter
Dutch Belted
Florida Scrub
Galloway
Guernsy
Kerry
Lincoln Red
Lineback (Gloucester and Witrik)
Milking Devon
Milking Shorthorn
Murray Grey
Red Poll
Scottish Highland
South Devon
Texas Longhorn
Welsh Black
White Park

Goats

African Pygmy
Nigerian Dwarfs
Spanish Angora (miniature)
Tennessee "Fainting"
Sable

Horses

American Cream Draft
Canadian
Chicksaw
Chunk Morgan
Cleveland Bay
Indian
Kabata
Missouri Fox Trotting
Norwegian Fjord
Spanish Jackstock
Spanish Mustang
Suffolk Punch
Welsh Cob

Sheep

Barbados
Black Welsh Mountain
Clun Forest
Cotswold
Florida Native
Hog Island
Jacob
Karakul

Kerry Hill
Leicester Longwool
Lincoln
Louisianna Native
Merino (Type A Vermont)
Morlams
Moufflon
Multinipple
Navajo
No Tail
North Country Cheviot
Otter
Scottish Blackface
Shetland
St. Croix
Teeswater
Wiltshire Horn

Swine

Brahma
Choctaw
Essex
Hereford
Large English Black
Mulefooted Hog
Ossabow
Piney Woods Rooter
Razor Back
Red Wattler
Tamworth

Associations

Cattle

Ayrshire
Ayrshire Breeders Association
J.D. Dodd, Executive Secretary
Brandon, VT 05733
(802)247-5774

Belted Galloway
The Belted Galloway Society
Summitville, OH 43962

Northeast Belted Galloways
A.H. Chatfield, President
Aldermerle Farm
Rockport, ME 04856

Dexter
American Dexter Cattle Association
P.O. Box 56
Decorah, IA 52101

Dutch Belted
Dutch Belted Cattle Association of America,
Inc.
Box 358
Venus, FL 33960

Galloway
American Galloway Breeders Association
1204 W. University
Suite 311
Denton, TX 76201

Lincoln Red
The Canadian Lincoln Red Association
Mrs Judy McDonald
Makwa Ranches
Box 644, Rocky Mountain House
Alberta, Canada TOM 1TO

Lineback (Gloucester and Witrik types)
American Lineback Registry
David Warden
Barnet, VT 05821

Milking Devon
American Milking Devon Association
John Wheelock, Secretary
Milton, VT 05468

Milking Shorthorn
American Milking Shorthorn Society
1722-JJS, Glenstone
Springfield, MO 65804

Murray Grey
American Murray Grey Association
1222 N. 27th St.
Billings, MT 59107

Red Poll
American Red Poll Association
Box 35519
Louisville, KY 40232

Scottish Highland
American Scottish Highland Breeders
Association
Box 81
Remer, MN 56672

Northeast Scotch Highland Association
Kathy Boone, Secretary
P.O. Box 502
Palmer, MA 01069
(413)283-9065

South Devon
North American South Devon Association
Dr T.E. Fitzpatrick,
Executive Secretary
Box 68, Lynnville, IA 50153
(515)527-2437

Texas Longhorn
Texas Longhorn Breeders Association of
America
3701 Airport Freeway
Ft. Worth, TX 76111

White Park Cattle
White Park Cattle Association of America
419 N. Water St.
Madrid, IA 50156

Horses

American Cream
American Cream Draft Horse Association
Karene Topp, Secretary
Route 1, Box 88
Hubbard, IA 50122

Chickasaw
Chickasaw Horse Association, Inc.
P.O. Box 607
Love Valley, N.C. 28677

Indian
American Indian Horse Registry
Route 1, Box 64
Lockhart, TX 78644

Spanish Mustang
Southwest Spanish Mustang Association,
Inc.
Gilbert H. Jones, Chairman
P.O. Box 148
Finley, OK 74543

Suffolk
American Suffolk Horse Association
M.M. Read, Secretary
15B Roden
Whichita Falls, TX 76311

Donkeys & Mules
The American Donkey and Mule Society,
Inc.
Route 5, Box 65
Denton, TX 76201

Sheep

Clun Forest
North American Clun Forest Association
High Meadow Farm, R.F.D. 2,
Box 100
Ferryville, WI 54628

Cotswold
American Cotswold Record Association
Pat Frisella, Secretary
282 Meaderboro Rd.
Rochester, N.H. 03867

Jacob
Multi-Horned Sheep Breeders Newsletter
Ingrid Painter, Editor
1607, 232 Ave., N.E.
Redmond, WA 98052
(202)855-3508

Karakul
American Karakul Fur Sheep Registry
8756 Concord Rd.
Box 22
Powell, OH 43065
(614)881-4130

Lincoln
National Lincoln Sheep Breeders
Association
Teresa Kruse, Secretary
Route 6, Box 24
Decatur, Illinois 62521

Navajo
Multi-Horned Sheep Breeders Newsletter
(see Jacobs above)

Swine

Hereford National Hereford Hog Record
Association
Route 1, Box 37
Flandreau, SD 57028

Mulefooted
R.M. Holliday
Route 2, Box 266
Louisianna, MO 63353

Tamworth
Tamworth Swine Association
Route 2, Box 36
Winchester, OH 45697

Appendix V
Places to Visit

Farms open to the public

Cotswold Farm Park
Guiting Power
Cheltenham
Glos GL54 5UG

Appleby Castle
Appleby-in-Westmorland
Cumbria CA16 6XH

Parke Rare Breeds Farm
Bovey Tracey
Devon

Riber Castle Fauna Reserve and Wildlife
Park
Riber Castle
Matlock
Derbyshire DE4 5JU

Tilgate Park Nature Centre
Crawley
West Sussex RH10 5PQ

Wimpole Home Farm
Wimpole
Arrington
Royston
Herts SG8 0BW

Croxteth Home Farm
Croxteth Hall
Liverpool L12 0HB

Graves Park
(City of Sheffield Recreation Dept)
PO Box 151
Meersbrook Park
Sheffield S8 9FL

Shugborough Park Farm
The Shugborough Estate
Milford
Stafford ST17 0XB

Great Hollander Farm
Mill Lane
Hildenborough
Kent

Ashdown Forest Farm
Wych Cross
East Sussex

Acton Scott Working Farm Museum
Wenlock Lodge
Acton Scott
Church Stretton
Shropshire

Kentwell Hall
Long Millford
Suffolk

Temple Newsome
Leeds

Parke Rare Breeds Farm
Bovey Tracey
Devon

Private breeding centres

Royal Agricultural Society of England
National Agricultural Centre
Stoneleigh
Kenilworth
Warwicks CV8 2LZ

Ash Farm (Mr & Mrs M.M. Rosenberg)
Iddesleigh
Winkleigh
Devon EX19 8SQ

Linga Holm
c/o RBST
4th Street
NAC
Stoneleigh

Ravendale Farm (Mr & Mrs J.H. Latimer)
Higham Gobion
Hexton
Hitchin
Herts SG5 3HS

Cluanie-an-Teanassie (Mr & Mrs
A. Crawford)
Beauly
Inverness

East Hele Farm (Mr & Mrs R.A. Petch)
Kingsnympton
Umberleigh
Devon EX37 9TB

East Johnston (Mrs E. Brown)
Bish Mill
South Molton
North Devon

Babylun Sheep (Mr & Mrs P.M. Mapson)
2 Curlew Close
St Ives
Huntingdon
Cambs PE17 4HL

Gillhouse Herd (Mr & Mrs W. Mills)
Zeal Monachorum Crediton
Devon

The Longhouse (Dr P. Wade-Martins)
Eastgate Street
North Elmham
Dereham
Norfolk NR20 5HD

Toddington Manor Farms (Sir Neville &
Lady Bowman-Shaw)
Toddington
Dunstable, Beds

Bibliography

Alderson, L.A., *A Chance to Survive*, David and Charles 1979
Cowerd, T.A., *The Fauna of Cheshire*, Wetherby and Co. 1910
Green, C.E., and Young, D., *Encyclopaedia of Agriculture*, William Green and Sons 1908
Hammond, Dr John, 'Polled Cattle' in *Endeavour* 1950
Henson, Elizabeth, *Rare Breeds in History* 1982
Leakey, Richard, and Lewin, Roger, *Origins*, MacDonald and Jane's 1977
Leigh, Charles, *Natural History of Lancashire and Cheshire*, Oxford 1700
Leonard, J.N., *The First Farmers*, Time Life Books 1974
Ormerod, George, *History of the County Palatine of Cheshire*, ed. T. Helsby, E.T. Morten 1980
Prentice, E. Parmalee, *Breeding Profitable Dairy Cattle*, Rich and Cowan 1935
Sherer, John, *Rural Life*, London Printing and Publishing Co. Ltd 1850
Spedding, C.R.W., *Elements of Agriculture*, ed. W. Fream, John Murray 1983
Storer, Rev. John, *The White Cattle of Great Britain*, Cassell, Petler and Gilpin 1877
Trow-Smith, Robert, *A History of British Livestock Husbndry*, Routledge and Kegan Paul 1959
Whitehead, G. Kenneth, *The Ancient White Cattle of Britain*, Faber and Faber 1953
Wilson, Rev. John M., *The Farmer's Dictionary*, A. Fullerton and Co.

Cymdeithas Defaid Lleyn 1982
Fact Sheets, Rare Breeds Survival Trust
Graves Park Animals, City of Sheffield Metropolitan District Recreation Department 1985
Shorthorn Breeders Guide, Council of Shorthorn Society of Great Britain and Ireland 1920
The Ark, Rare Breeds Survival Trust

Further Reading

The Ark, the monthly magazine of the Rare Breeds Survival Trust
Alderson, Lawrence, *The Chance to Survive*, David and Charles 1979
Alderson, Lawrence, *Rare Breeds*, Shire Publications 1984
Bailey, Eric, *The Domestic Poultry Keeper*, Blandford Press 1985
Clutton-Brock, Juliet, *Domesticated Animals from Early Times* (no date)
Friend, John, *Cattle of the World*, Blandford Press 1978
Henson, Elizabeth, *Rare Breeds in History* 1982
Leonard, J.N., *The First Farmers*, Time-Life Books 1974
Ponting, Kenneth, *Sheep of the World*, Blandford Press 1980
Spedding, C.R.W., *Elements of Agriculture* (Fream), John Murray 1983
Trow-Smith, Robert, *A History of British Livestock Husbandry* (2 Vols), Routledge and Kegan Paul 1959
Vandivert, Rita, *To the Rescue*, Frederick Warne (no date)
Whitehead, G. Kenneth, *The Ancient White Cattle of Britain*, Faber and Faber 1953

Index

Figures in italics refer to page numbers of illustrations.